B.C. Science
PROBE
7

Nelson

B.C. Science PROBE 7

AUTHORS

Anita Chapman
Educational Consultant and Author,
Chapman & Associates Educational Consulting, Inc.

David Barnum
Sunshine Coast School District, #46

Carmen Dawkins
North Okanagan-Shuswap School District, #83

William Shaw
Educational Consultant/Teacher and Author

PROGRAM CONSULTANT

Arnold Toutant
Educational Consultant, A. Toutant Consulting Group Ltd.

THOMSON

NELSON

Australia   Canada   Mexico   Singapore   Spain   United Kingdom   United States

THOMSON

NELSON

B.C. Science Probe 7

Authors
Anita Chapman
David Barnum
Carmen Dawkins
William Shaw

Program Consultant
Arnold Toutant

Associate Vice President of Publishing:
David Steele

**Senior Publisher,
Assessment and Science:**
Bill Allan

Acquisitions Editor, Science:
John Yip-Chuck

**Executive Managing Editor,
Development:**
Cheryl Turner

Program Manager:
Lee Geller

Project Editor:
Lee Ensor

Developmental Editors:
Lee Ensor, Janis Barr

Editorial Assistant:
Alisa Yampolsky

Executive Managing Editor, Production:
Nicola Balfour

Senior Production Editor:
Deborah Lonergan

Copy Editor:
Paula Pettitt-Townsend

Proofreaders:
Paula Pettitt-Townsend, Elizabeth
Salomons, Gilda Mekler

Indexer:
Noeline Bridge

Senior Production Coordinator:
Sharon Latta Paterson

Creative Director:
Angela Cluer

Text Design:
Kyle Gell Design, Peter Papayanakis,
Ken Phipps

Art Management:
Allan Moon, Suzanne Peden

Composition Team:
Kyle Gell, Allan Moon

Cover Design:
Ken Phipps

Cover Image:
Photodisc Blue/Photodisc
Collection/Getty Images

Illustrators:
Steven Corrigan
Deborah Crowle
Margo Davies LeClair
Kyle Gell
Irma Ikonen
Imagineering
Kathy Karakasidis
Dave Mazierski
Dave McKay
Allan Moon
Bart Vallecoccia
Cynthia Watade
Dave Whamond

Photo Researcher:
Karen Becker

Permissions:
Karen Becker

Printer:
Transcontinental Printing Inc.

**Library and Archives Canada
Cataloguing in Publication Data**

B.C. science probe 7 / Anita Chapman ...
[et al.].

Includes index.
ISBN 0-17-627184-8

1. Science—Texbooks. I. Chapman,
Anita II. Title: B.C. science probe seven.

Q161.2.B39 2005   500   C2004-
905910-6

REVIEWERS

Aboriginal Education Consultant
Mary-Anne Smirle
Peachland Elementary School, Central Okanagan
  School District (#23), B.C.

Accuracy Reviewers
Anne Laite
Chatelech Secondary School, Sunshine Coast
  School District (#46), B.C.

James A. Hebden, Ph.D.
Formerly of Kamloops/Thompson School
  District (#73), B.C.

Brian Herrin
Simon Fraser University, B.C.

Assessment Consultant
Anita Chapman
Educational Consultant and Author, Chapman &
  Associates Educational Consulting, Inc.

ESL Consultant
Vicki McCarthy, Ph. D.
Dr. George M. Weir Elementary School, Vancouver
  School District (#39), B.C.

Literacy Consultant
Sharon Jeroski
Educational Consultant and Author,
  Horizon Research and Evaluation, Inc.

Numeracy Consultant
Bob Belcher
Happy Valley at Metchosin Elementary School,
  Sooke School District (#62), B.C.

Professional Development Consultant
Brian Herrin
Simon Fraser University, B.C.

Safety Consultant
Marianne Larsen
Formerly of Sunshine Coast School
  District (#46), B.C.,
  Greater Vancouver Regional Science Fair

Technology Consultant
Al Mouner
Millstream Elementary School, Sooke School
  District (#62), B.C.

REVIEWERS

Advisory Panel and Teacher Reviewers

Doug Adler
Department of Curriculum Studies, Science Education, University of British Columbia

Wade Blake
Rutherford Elementary School, Nanaimo-Ladysmith School District (#68), B.C.

Jim Chong
Westerman Elementary School, Surrey School District (#36), B.C.

Burt Deeter
James Ardiel Elementary School, Surrey School District (#36), B.C.

Pat Horstead
Mount Crescent Elementary School, Maple Ridge-Pitt Meadows School District (#42), B.C.

Jillian Lewis
Lyndhurst Elementary School, Burnaby School District (#41), B.C.

Susan Martin
Cliff Drive Elementary School, Delta School District (#37), B.C.

Ann McDonnell
Blewett School, Kootenay Lake School District (#8), B.C.

Darlene Monkman
Formerly of Nanaimo-Ladysmith School District (#68), B.C.

Karen Morley
North Surrey Learning Centre, Surrey School District (#36), B.C.

Noreen Morris
Trafalgar Elementary School, Vancouver School District (#39), B.C.

Mary-Anne Smirle
Peachland Elementary School, Central Okanagan School District (#23), B.C.

Heather Stannard
Khowhemun Elementary School, Cowichan Valley School District (#79), B.C.

Michelle Stecher
Brookside Elementary School, Surrey School District (#36), B.C.

Patricia Tracey
Formerly of Abbotsford School District (#34), B.C., Fraser Valley Regional Science Fair Foundation

Kyme Wegrich
Glen Elementary School, Coquitlam School District (#43), B.C.

Stan Yuen
Maillard Middle School, Coquitlam School District (#43), B.C.

Acknowledgements

The authors and publisher gratefully acknowledge the contributions of the following schools who participated in photo shoots: Cliff Drive Elementary School, Delta, B.C. (SD #37), and Constable Neil Bruce Middle School, Kelowna, B.C. (SD #23).

CONTENTS

UNIT B: CHEMISTRY

UNIT C: EARTH'S CRUST

EXPANDING THE WORLD OF SCIENCE

What is science? Do all cultures have science? How would you explain what "science" is to someone who is unfamiliar with the term?

When students your age are asked to explain what science is, quite often a picture pops into their minds. Some students see a scientist in a white lab coat, working alone in his or her laboratory with lots of specimens in jars and potions bubbling out of beakers. Some students think of science as microscopes and telescopes. Others think that science means experiments to be done and formulas to be learned.

The dictionary says that science is what we know about the physical and natural world, for example, facts, laws, and relationships. Science is knowledge that is learned by careful observation and experimentation. For example, we know that all living things are made up of cells. This is a scientific fact. This knowledge came from scientists carefully observing living things.

People often think that the only important science is the kind that is taught in classrooms. There is another kind of science, however. This science, sometimes known as **Indigenous Knowledge** (IK) or Traditional Ecological Knowledge (TEK), can give us valuable information. This knowledge can help us learn better ways to live in our world.

First Nations and Inuit peoples have lived in their traditional territories long before the first explorers and immigrants arrived in North America. These Indigenous groups, and later the Metis people, lived very closely with nature. They developed very detailed knowledge about the places they lived and about their environment. This includes knowledge about plants, animals, weather, and landforms. The information has been carefully passed on from one generation to the next so that the knowledge won't be lost. In each community, the Elders are the people who know this important information and who teach it to the young people so that they can live carefully and successfully in their environment.

For centuries, and continuing today, Indigenous peoples around the world carefully observe, describe, explain, predict, and work with their natural world. Scientists also observe, describe, predict, and work with the natural and physical world. Both of these groups have much to learn from each other.

In British Columbia, like other places all over the world, more and more scientists are asking to work with and learn from Aboriginal peoples and communities. When this happens it can be a powerful partnership for everyone involved. It is important for Aboriginal peoples that their ways are respected and valued by others. It is important for all of us.

Mary-Anne Smirle
Metis Nation
L'Hirondelle clan

UNIT A

ECOSYSTEMS

CHAPTER
1 Ecosystems support life.

CHAPTER
2 Energy flows and matter cycles in ecosystems.

CHAPTER
3 Human survival depends on sustainable ecosystems.

Preview

Have you heard any recent news about the environment? Does there seem to be a lot of bad news about pollution, endangered species, and global warming? Do you ever wonder whether you should worry about these things? Could these things really affect your life?

No organism lives alone. On Earth, every living thing interacts with other living things and with the non-living parts of its environment. Your survival, and the survival of every other living thing, depends on these interactions.

In this unit, you will learn how living things affect their surroundings and are affected by their surroundings. You will learn how plants and animals, including humans, depend on one another and how they interact with their environment. By learning about these interactions, you will discover how you fit into the living world.

TRY THIS: RECORD ENCOUNTERS WITH ORGANISMS

Skills Focus: observing, inferring

With a partner, record all the different organisms you have encountered over the past week. Draw a web and use arrows to show which organisms may have been interacting with each other. Did you encounter any dead or decaying organisms? Do you think you encountered any microscopic organisms? Remember to include yourself in your web, as well as the food you have eaten.

By the time you finish your web, it may look like a spider's web. Join with another pair of students to compare your webs. How many organisms do you have in common? Did they identify the same interactions you did or different interactions?

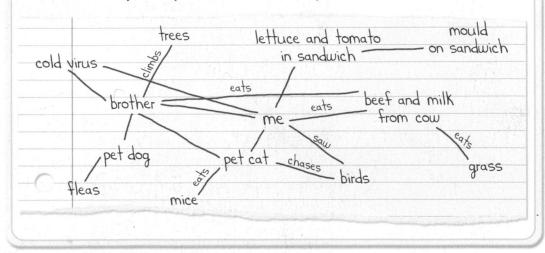

The Khutzeymateen Valley, near Prince Rupert, British Columbia.

· CHAPTER ·

1

Ecosystems support life.

KEY IDEAS

▶ Ecosystems are made up of living and non-living things.

▶ Groups of living things interact within ecosystems.

▶ All the ecosystems on Earth are interconnected.

▶ Limiting factors determine which species' needs will be met in an ecosystem.

▶ Living things interact in different ways.

▷ **LEARNING TIP**

As you read the first two paragraphs, try to answer the questions using what you already know.

How large is the world of this bee? What does it need in its environment in order to survive? How is the bee affected by other living things? How does it affect other living things? How might the bee affect your life? Can it affect your life even if it lives hundreds of kilometres away?

How large is your world? How are the things you need to survive like the things the bee needs to survive? How do other living things in your environment affect you? How do you affect them? How do your actions affect the bee's chances of survival? Can your actions still affect the bee if you live hundreds of kilometres away?

In this chapter, you will learn about the connections among living things. You will also learn about the connections between living things and their non-living environment.

What Is an Ecosystem?

The Khutzeymateen [K'TZIM-a-deen] Valley (**Figure 1**) is a large, undisturbed area of wilderness in one of British Columbia's coastal rain forests. The valley is a traditional hunting and fishing area for the Gitsiis people. It is an area of high rainfall, with rugged mountains, creeks, and a large river than runs down to the ocean. "Khutzeymateen" is a Tsimshian [SIM-she-an] word that means "a confined space for salmon and bears." The Khutzeymateen Valley is home to more than 50 grizzly bears. In 1994, it became the first grizzly bear sanctuary (protected area) in Canada. It is also home to salmon, beavers, wolves, otters, birds, insects, trees, shrubs, and many other living things. All these living things depend on the environment for survival.

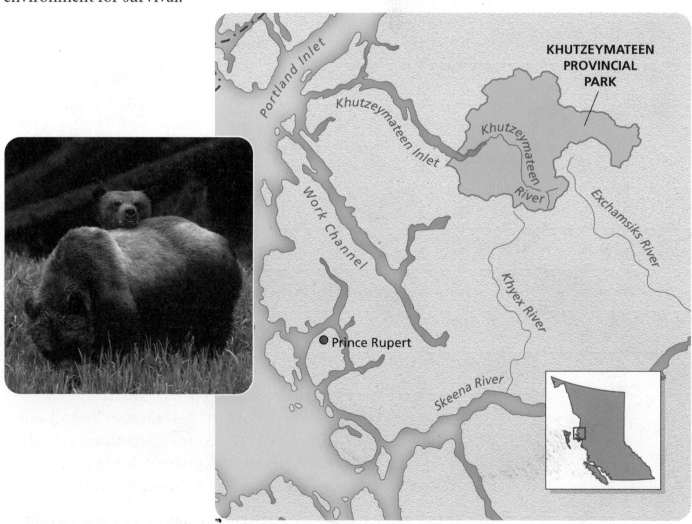

Figure 1
Khutzeymateen Provincial Park provides a protected area for grizzly bears.

▷ **LEARNING TIP**

Important vocabulary words are highlighted. These are words you should learn and use when you answer questions. These words are also defined in the glossary at the back of this book.

▷ **LEARNING TIP**

It is easier to remember scientific terms if you understand the root words. The Greek word *micro* means "small." Therefore, micro-organisms are simply small organisms. Can you think of other science words that start with "micro"?

The Living Environment

The Khutzeymateen Valley has both living and non-living parts. The living parts, such as plants and animals, are called **organisms.** Some of the organisms, such as bacteria and the tiniest algae, are too small to be seen with your eyes only. Organisms that are too small to be seen without the help of a microscope are called **micro-organisms.**

Each different type of organism—plant, animal, or micro-organism—is known as a species. Grizzly bears are a **species** (**Figure 2(a)**). All the members of one particular species in a given area, such as the Khutzeymateen Valley, are called a **population.** For example, all the grizzly bears in the Khutzeymateen Valley form a population (**Figure 2(b)**). When two or more populations of different species live in the same area, they form a **community** (**Figure 2(c)**). The community in the Khutzeymateen Valley includes populations of grizzly bears, coho salmon, red elderberry, Sitka spruce, and ravens.

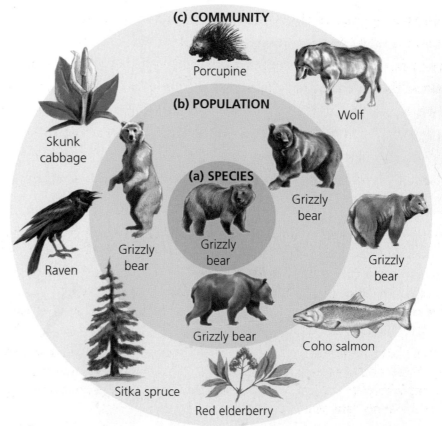

Figure 2
A nested circle diagram shows how parts fit into a whole. Each species is part of a population, and each population is part of a community.

The Non-Living Environment

The non-living parts of the Khutzeymateen Valley include the sunlight, rain and snow, soil, creeks and rivers, mountains, and temperature (**Figure 3**). These non-living parts of the environment provide many of the things that the organisms need to survive. Plants need soil, water, and sunlight. Animals need water, shelter, and an appropriate temperature range.

Figure 3
What non-living parts of the Khutzeymateen Valley can you identify in this photo?

The living parts of the Khutzeymateen Valley interact with each other and with the non-living parts of their environment. A grizzly bear eating red elderberries is an interaction between two living parts of the environment. Rain washing away soil is an interaction between two non-living parts of the environment. Sitka spruce trees using sunlight to grow is an interaction between a living part of the environment and a non-living part. The network of interactions that link the living and non-living parts of an environment is called an **ecosystem.**

CHECK YOUR UNDERSTANDING

1. List some living parts of the Khutzeymateen Valley on one side of a page in your notebook. List some non-living parts on the other side of the page. Draw lines to show interactions between the living and non-living parts of the ecosystem.
2. Choose a wild animal species in the Khutzeymateen Valley ecosystem. Draw and label a nested circle diagram like **Figure 2** to show this species in its population and community. Label your diagram using the terms "species," "population," and "community."
3. Describe an interaction between two living parts of the environment and two non-living parts.

LEARNING TIP

Do not guess. Look back through the section to find the answers. Even if you remember the answer, it is a good idea to go back and check it.

● SKILLS MENU

○ Questioning   ● Observing
○ Predicting   ○ Measuring
○ Hypothesizing   ● Classifying
○ Designing Experiments   ● Inferring
○ Controlling Variables   ○ Interpreting Data
○ Creating Models   ● Communicating

▷ **LEARNING TIP**

It is not always a good idea to work quickly. Do not rush when you are making your observations. The longer you observe, the more you will notice. Learning slowly sometimes results in learning more. "Slow knowledge" is usually more detailed and complete.

Your Schoolyard Ecosystem

People who live in an area for a long time can get to know the environment very well. For example, many Aboriginal peoples have developed very detailed knowledge of their "home places." The first step in developing this knowledge is observation.

In this investigation, you will observe both the living and non-living parts of the environment near your school (**Figure 1**). You will observe some of the interactions that are taking place, and record your observations. As well, you will infer, or figure out from your observations, other interactions that you cannot see.

Figure 1

If you observe carefully, you will find many living and non-living things, even in an empty-looking schoolyard like this.

Question

Are there clear examples of interactions among the living and non-living parts of your schoolyard environment?

Materials

drawing paper

coloured pencils

hand lens

field guides

- large sheet of drawing paper
- coloured pencils
- hand lens
- field guides

✋ Be aware of any allergies you may have to plants and animals. Watch for poisonous plants or animals that are found in the area. Hand lenses can concentrate the Sun's energy and start a fire. Do not leave a hand lens on dry grass.

Procedure

1 Choose a suitable area of your schoolyard to study. The area should be about 1 m by 1 m. If you are not able to use your schoolyard, use a nearby park, vacant lot, or field.

2 In your notebook, make a table like the one below.

3 Walk slowly around your study area to look for organisms. Remember to look for dead organisms (such as fallen logs) and signs of organisms (such as empty shells or feathers), as well. Record your observations in the first column of your table.

Note: Do not pick or break any plants or damage any flowerbeds. If you turn over rock or log to see what is underneath, be sure to replace the rock or log exactly the way you found it.

4 For each organism, record any connections to other living things in the second column of your observation table. For example, a plant may have small insects living on it. A spider may have the remains of its food in its web. Look closely, and use your hand lens.

5 In the third column, note any connections between each organism and the non-living parts of the environment. For example, is the soil sandy or is it hard clay? Is the area wet or dry? If the area is wet, where did the water come from?

6 If you have time, use your field guides to identify and learn about any organisms that are not familiar to you.

Schoolyard Observations

Organisms	Connections to other living things	Connections to non-living parts of the environments
earthworm		under rock
leaf	small holes – probably eaten by insects	
bird, mostly brown, small	perched in tree	drinks water in puddle, stays in shade
black ants, 2mm long	carrying insect wing	go down cracks in ground

Analyze and Evaluate

1. Make a class list of all the species of organisms that were observed during this schoolyard study. Remember to include yourself. This is the community of living things that share your schoolyard with you.

2. Choose one animal or one plant to analyze in more detail. Draw the animal or plant in the centre of a large sheet of unlined paper, leaving plenty of space around your drawing. In the space, write several living and non-living parts of the environment that might affect your organism or be affected by it. Use your observations for this information, and ask your classmates if they have any additional observations. Use two different colours: one for the living parts and the other for the non-living parts. Use a line to connect each part of the environment to the organism at the centre. Identify the interaction along the line (**Figure 2**).

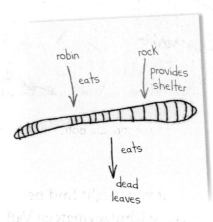

Figure 2
Identify the connections between the living and non-living parts of this worm and its environment.

Apply and Extend

3. Study your drawing of a plant or animal and its interactions. What might happen to the organism if you removed one of the other living parts of the environment? What might happen if you removed one of the non-living parts of the environment? Explain by describing the interactions.

▐▶ CHECK YOUR UNDERSTANDING

1. Explain why you might get different results if you did your study
   - at night
   - in another season
2. You can sometimes "read between the lines" of your observations. Based on what you already know, you can suggest more things than you actually observed directly. We call this inferring. Give three examples of interactions that you did not see directly but were able to infer.

▷ **LEARNING TIP**

For help with inferring, see "Inferring" in the Skills Handbook.

Look at **Figure 1**. What do the two photos have in common?

Figure 1
A discarded bottle with rainwater in it, and a backyard koi pond, are both examples of small ecosystems.

Both photos depict small ecosystems that you might find near your home. Ecosystems can be as large as the Khutzeymateen Valley or as small as a discarded bottle or a koi pond. Ecosystems can be created or altered by humans, or they can be more natural, such as a wilderness area.

When you think about ecosystems, such as the koi pond and the Khutzeymateen Valley, keep in mind that an ecosystem is not really a place. It is a set of interactions among the living and non-living parts of the environment. Also keep in mind that there are ecosystems within ecosystems.

You could go to the Khutzeymateen Valley and study the interactions that occur in one rotting log, along one creek, or in the whole valley (**Figure 2**). In each of these ecosystems, you would find living and non-living parts interacting with each other. Although studying a small ecosystem is often more practical and convenient, you should never forget that it is part of a larger ecosystem.

Khutzeymateen Valley Ecosystem

Carm Creek Ecosystem

Rotting Log Ecosystem

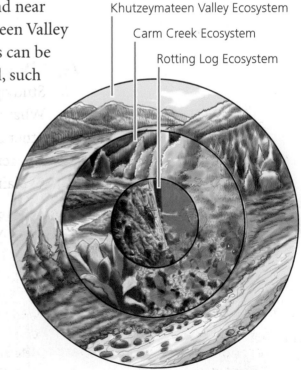

Figure 2
Nested circles are a good model for showing how smaller ecosystems are nested within larger ecosystems.

The whole Earth is one large ecosystem called the **biosphere.** The biosphere includes all the places on Earth where living things are found, from mountaintops to the deepest parts of the oceans. Since it is difficult to study all the interactions in such a large ecosystem, scientists divide the biosphere in various ways.

Scientists have identified large areas of Earth that have roughly the same temperatures and the same amounts of rain or snow. These large areas are called **biomes** (**Figure 3**).

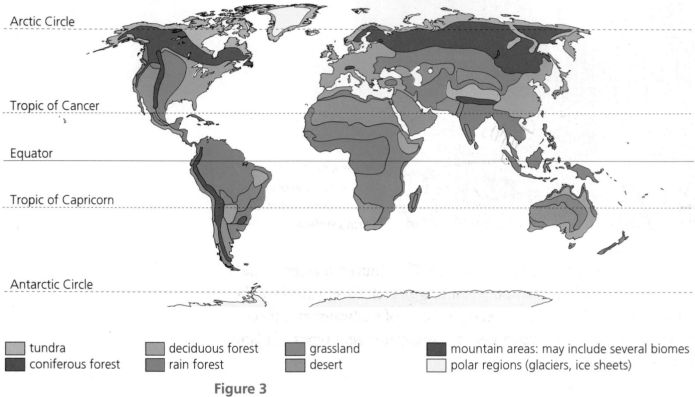

Arctic Circle

Tropic of Cancer

Equator

Tropic of Capricorn

Antarctic Circle

- tundra
- coniferous forest
- deciduous forest
- rain forest
- grassland
- desert
- mountain areas: may include several biomes
- polar regions (glaciers, ice sheets)

Figure 3
This map shows the major biomes of the world. Each biome is named after the most common type of vegetation that is found there.

Biomes are still very large, so scientists often study smaller ecosystems within each biome. All these ecosystems overlap. They are connected to each other and to the ecosystems in the oceans.

Together, the world's oceans form one vast ecosystem (**Figure 4**) that is connected to the ecosystems on land. Just as there are ecosystems within ecosystems on land, there are many ecosystems within the vast ecosystem of the oceans. For example, smaller ocean ecosystems along the coasts are very different from ecosystems in the open ocean. Ecosystems near the surface of the ocean are very different from ecosystems in the deep.

Figure 4
This is a photo from space. Notice that the oceans cover almost three-quarters of Earth's surface.

Land and ocean ecosystems overlap. The Khutzeymateen estuary is a lush, green place where the Khutzeymateen River runs into the Pacific Ocean (**Figure 5**). Here, the mixing of salt water and fresh water creates a unique ecosystem for many different types of plants and animals.

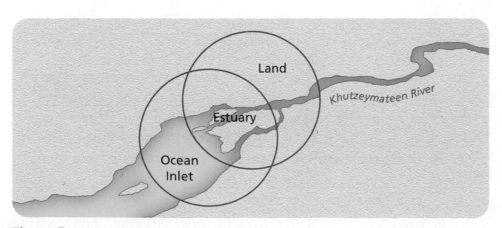

Figure 5
A Venn diagram can be used to show how ocean and land ecosystems overlap.

1.3 Ecosystems Within Ecosystems

Land and ocean ecosystems are linked where they meet. They are also linked over great distances through the movement of animals and the flow of water. For example, western grebes (**Figure 6**) spend their winters on the coastal waters of British Columbia, yet fly back to large interior lakes each spring to reproduce and rear their young.

Figure 6
An interior lake with western grebes

In many parts of British Columbia, salmon link ocean and land ecosystems. Salmon that hatch in inland rivers spend their adult lives in the open ocean. Then they swim hundreds of kilometres, back to the rivers where they hatched, to spawn. Even if you live far inland, rivers and streams that flow through or near your community connect you to the ocean.

Dividing the biosphere into smaller and smaller ecosystems helps us study them in more detail. It is important, however, to remember that all of Earth's ecosystems are connected and affect one another.

⫸ CHECK YOUR UNDERSTANDING

1. The Khutzeymateen Valley and a koi pond are both examples of ecosystems. How are they different? How are they the same?

2. List three small ecosystems that you might find within the larger ecosystem of your schoolyard.

3. In what biome do you live? What type of vegetation is this biome named after?

4. Use a nested circle diagram to show how your schoolyard ecosystem is part of the biosphere. Compare your diagram with your classmates' diagrams. Do all the diagrams have the same number of circles? Explain how there can be several different correct diagrams.

▷ **LEARNING TIP**

For help in how to use a nested circle diagram to show parts within a whole, see "Using Graphic Organizers" in the Skills Handbook.

The Needs of Living Things

1.4

Organisms must have their basic needs (such as food, water, and suitable living conditions) met in order to survive. If an ecosystem does not supply an organism with its basic needs, the organism will not be found in this ecosystem. Think of what you need to survive. Plants and animals need these things too. Like you, if they have the right combination of all of these things, they will probably thrive.

Survival Needs

The survival needs of plants and animals include the following:

- *Sunlight:* Plants need sunlight in order to produce food. As well, sunlight provides heat for both plants and animals.

- *Food:* Plants can produce their own food using sunlight, but animals must eat plants or other animals to get their food.

- *Air:* Animals need oxygen from the air. Plants need carbon dioxide from the air to make food with the help of sunlight.

- *Water:* The bodies of both plants and animals are mostly water. In fact, you can think of living things as sacs of water. Water has many important functions in the bodies of living things.

- *Shelter:* Some animals find natural shelter in their environment. Others, like beavers and wasps, build shelters using materials from their environment.

The physical space where a certain species lives is called its **habitat.** A species of plant or animal can only live in a habitat where its survival needs are met. Although most species need the same basic things, the amount and type they need may be very different. Different ecosystems provide different amounts of sunlight and water and different types of shelter. For example, the Khutzeymateen Valley ecosystem gets about 304 cm of rainfall and has about 1400 h of sunlight per year. The antelope brush ecosystem, in the south Okanagan, gets about 34 cm of rainfall and has about 2000 h of sunlight per year (**Figure 1**).

Make connections to your prior knowledge. What do you already know about survival needs from previous grades? Is there any new information here?

Figure 1
Many plants and animals that thrive in the rain forest ecosystem of the Khutzeymateen Valley could not survive in the antelope brush ecosystem of the south Okanagan because their survival needs would not be met.

▷ **LEARNING TIP**

Before reading the next four pages, "walk" through them and note the subheadings. Make a list of the limiting factors you expect to learn about.

Limiting Factors in the Non-Living Environment

All organisms have basic survival needs. If one of these needs is not met in an ecosystem, then the organism will not be able to live there. Any part of the non-living environment that determines whether or not an organism can survive is called a limiting factor. Limiting factors include physical barriers, sunlight, water, temperature, and soil.

Physical Barriers

Often an organism is not found in a particular ecosystem simply because it is unable to get to the ecosystem. Oceans, rivers, mountain ranges, and other landforms can block a plant or animal from moving to another suitable area. These landforms are called physical barriers.

Gwaii Haanas [G-why Hah-nas], the group of islands at the south end of Haida Gwaii [HY-duh G-why], has 39 plant and animal species that are not found anywhere else in the world. One of these species is the Haida Gwaii black bear, the largest black bear in North America (**Figure 2**). Many of the organisms in Gwaii Haanas would be able to live in similar habitats on the mainland, but it is too far for seeds to travel or animals to swim.

Figure 2
The Haida Gwaii black bear is only found in Gwaii Haanas because the stretch of water between the islands and the mainland is a physical barrier.

Sunlight

The amount of sunlight in an area can determine whether or not an organism can live there. Some plants grow better in bright sunlight. Other plants grow better in shade. For example, dandelions grow better in sunny places, but skunk cabbages grow better in shade.

In water ecosystems, sunlight can only shine down to a certain depth. Organisms that use sunlight to produce food can only exist in areas close to the surface, which have enough sunlight (**Figure 3**).

Many of the reptiles that live in British Columbia, such as snakes and turtles, bask in the sunlight to raise their body temperatures (**Figure 4**). Reptiles that live in hot climates seek the shade to escape the heat.

Figure 3
Kelp requires sunlight to produce food.

Figure 4
A western painted turtle basks in the sunlight to warm itself.

Water

All organisms need water. How much water they need, when they need it, and what type of water they need (fresh or salty) varies, however. The availability and type of water in an ecosystem determines what organisms can live there.

Some plants and animals need to absorb or drink water every day. Other plants and animals can exist for a long time without water. Some animals live on land but need water to reproduce (**Figure 5**).

Figure 5
This Pacific Treefrog lives on land but returns to a pond to breed.

Most plants and animals need either salt water or fresh water. A few organisms, such as salmon, live in fresh water for one part of their life cycle and salt water for the other part. Some organisms, such as the Nootka rose and shooting star (**Figure 6**), may be found in estuaries where salt water and fresh water mix.

Figure 6
This shooting star can grow in very salty soils.

Temperature

Temperature can limit the survival of an organism, if the temperature is too hot or too cold for an extended period of time. In British Columbia, the temperature is usually the coldest in the north or on mountaintops. As you travel north or up mountains, you find a tree line (**Figure 7**). Above the tree line, the temperature is too cold for trees to grow.

Figure 7
Temperature is a limiting factor for the growth of trees.

Even short-term changes in temperature can affect survival. For example, the upper limit of water temperature for successful hatching of salmon eggs is 20°C.

Soil

In nature, soil gets its nutrients from the decomposition of plants and animals. The soil of the Khutzeymateen Valley has large amounts of broken-down plant material in it. Therefore, it holds water like a sponge. The soil in the antelope brush ecosystem contains very little plant material, so water runs through it very quickly. Plants need different types of soil. For example, Indian hellebore grows well in rich, moist soil, but sagebrush requires thin, dry soil (**Figure 8**).

Figure 8
Indian hellebore (left) and sagebrush (right) require different types of soil to grow well.

TRY THIS: *IDENTIFY THE BEST LIVING CONDITIONS*

Skills Focus: inferring, classifying

The plant tags in **Figure 9** describe the best living conditions for the different plants. You can infer the limiting factors for the plants from their tags.

Design a "best living conditions" tag for an organism. The organism could be you, a pet, or a plant or animal from the ecosystem in which you live.

Figure 9

▐▶ CHECK YOUR *UNDERSTANDING*

1. List the survival needs of all living things.

2. A gardener places a plant in a garden where there is a suitable amount of space and water, and suitable soil and temperature. The plant soon dies, however, because it is in the shade. What is the limiting factor for the success of this plant?

3. Logging companies are no longer allowed to remove trees that shade salmon streams. What limiting factor would exist for salmon if the trees were removed?

○ SKILLS MENU

● Questioning ● Observing
● Predicting   ● Measuring
● Hypothesizing ○ Classifying
● Designing   ● Inferring
  Experiments
● Controlling ● Interpreting
  Variables   Data
○ Creating   ● Communicating
  Models

Factors That Limit Yeast Growth

Did you know that yeast is a living thing (**Figure 1**)? It feeds, reproduces, and even excretes waste. Yeast lives in a moist, sugar-rich environment. In order to grow, it requires food in the form of a sugar. What are the limiting factors for yeast? Can it grow in any conditions? Does it grow better in certain conditions? Design an experiment to answer your own question about the growth of yeast.

 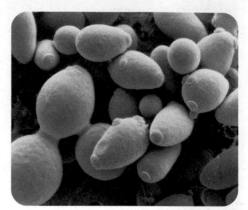

Figure 1
Baker's yeast grows and reproduces in the right conditions.

Question

Write a question you would like answered about factors that affect the growth of yeast. Think carefully about what you want to find out in your experiment. What do you want to know about yeast growth and limiting factors?

▷ **LEARNING TIP**

For help with writing a hypothesis, controlling variables, or writing up your experiment, see "Hypothesizing," "Controlling Variables," and "Writing a Lab Report" in the Skills Handbook.

Hypothesis

Write a hypothesis based on your question. Your hypothesis should be a cause and effect statement that you can test. It should be a sentence in the form "If … then …."

Materials

Make a point-form list of all the materials you will need to complete your experiment. Be sure to include exact sizes and quantities.

- Design a procedure for your experiment. In your design, include
  - descriptions of the independent, dependent, and controlled variables
  - a step-by-step description of the procedure
  - a list of safety precautions
- Be sure to include at least two controlled variables (variables that you will control during the experiment). Remember that you need to design a fair test.
- Submit your procedure, including any safety precautions, and a diagram of how you will set up the equipment to your teacher for approval. Your diagram should be at least a half page in size.

Data and Observations

Design a data table and record your observations.

Analysis

Look at your observation table. Why do you think you observed what you did? Do you think your experiment was affected by a variable that you did not control? If so, what might this variable have been?

Conclusion

Write a conclusion that explains the results of your experiment. Your conclusion should refer back to your hypothesis. Was your hypothesis correct, partly correct, or incorrect? Explain how you arrived at your conclusion.

Applications

What other questions could you ask about limiting factors for the growth of yeast?

LEARNING TIP ◁

You might graph your results in order to find a pattern. For help in graphing data, see "Using Graphic Organizers" in the Skills Handbook.

▶ **CHECK YOUR UNDERSTANDING**

1. Explain what conducting a fair test means. List two things in this experiment that contributed to it being a fair test.
2. Describe one thing that someone might do to make this an unfair test.

1.6 The Interactions of Living Things in Ecosystems

You have learned that the number of plants or animals in a population can be limited by factors in the non-living environment, such as sunlight and water. You have also learned that plants and animals interact with other living things. These interactions can place limits on population growth as well.

Competition

When you run a race or play a baseball game, you are competing. You are hoping to be more successful than your competitors. Other animals and plants also compete, often for life or death. Competition in ecosystems occurs when an organism tries to get what it needs to survive, but other organisms need and try to get the same things (**Figure 1**). For example, plants that grow close together all try to get water, sunlight, and nutrients from the same small area. They may all be small and thin, until some die and make more water, sunlight, and nutrients available for the remaining plants (**Figure 2**).

Figure 1
Which animals are competing for the salmon in this carving by Fred Davis (Haida)?

Figure 2
Which plants are winning this competition?

A competition can leave you feeling tired and weak. Plants and animals are also weakened by competition. It is easier for a disease to affect a weakened organism.

The competition for the resources in an ecosystem limits the sizes of populations. For example, grizzly bears compete with other grizzly bears for food and places to live. Each grizzly bear requires a very large area for gathering food. Even the huge Khutzeymateen Valley can only support about 50 grizzly bears.

Predator–Prey Interactions

An animal that hunts another living thing for food is called a **predator.** The organism that is being hunted is called the **prey.** A lynx (the predator) eating a snowshoe hare (the prey) is an example of a predator–prey relationship (**Figure 3**).

Figure 3
A lynx (the predator) is chasing a snowshoe hare (its prey).

A population of predators cannot increase unless there is enough prey. At the same time, the predators tend to keep the population of prey from increasing. As a result, there is usually a balance between predators and prey in an ecosystem. This balance is more like a teeter-totter than a level beam, with more prey or more predators at different times.

Case Study: The Search for an Explanation

The population of snowshoe hares in the Yukon Territory and other parts of northern Canada rises and falls in a cycle that is about 10 years long. Some Aboriginal peoples have known about this cycle for thousands of years. The hare were a major food source for the Aboriginal peoples. When the population of hares was at its lowest, they often went hungry.

Fur traders for the Hudson Bay Company also noticed this cycle. They bought the pelts of both the hare and one of its predators, the lynx. In 1925, a scientist graphed the Hudson Bay Company data (**Figure 4**). The graph showed that the lynx population also cycles in a pattern. The lynx pattern follows the hare pattern by about a year.

▷ **LEARNING TIP**

Look at each axis of the graph and the legend. What do they tell you about what the graph shows? Check your understanding by explaining the graph to someone else.

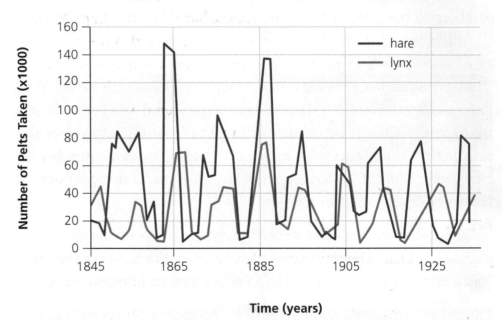

Figure 4

The size of the hare (prey) population is the main factor that controls the size of the lynx (predator) population.

At first, scientists thought that they had found a simple cause-and-effect relationship. If there were many snowshoe hares, then the lynx population would increase. When there were more lynx killing and eating hares, the hare population would decrease. As the hare population decreased, there was less food for the lynx and their population would also decrease. With fewer lynx killing hares, the hare population would increase again, and so on.

Scientists still think that the size of the hare population is the main factor that controls the lynx population. Although lynx eat other animals, such as grouse, they depend mainly on hares for food.

Scientists no longer think, however, that the lynx population is the main factor that controls the hare population. Snowshoe hare populations cycle even in areas where there are no lynx. This may be due to the fact that hares have several other predators, including wolves, owls, and humans.

LEARNING TIP

If you find this explanation difficult, read more slowly at the beginning until you feel you understand the content.

Some scientists thought that perhaps the hares run out of food when their numbers are high. To find out, they fenced areas of plants to keep the hares out. To their surprise, they found little difference between the number of plants in the fenced areas and the number of plants in the unfenced areas where the hares were feeding. When scientists looked at the plants, however, they found that the plants nibbled by hares had produced substances that made them less tasty. The hares were not running out of food. The food was there, but the hares could not eat it. The scientists had discovered another cycle. As the number of hares increases, the number of plants that produce the unappetizing substances also increases. This decreases the number of hares, which, in turn, decreases the number of plants that produce the unappetizing substances.

Today, scientists are still testing hypotheses to explain the hare population cycle. They still cannot fully explain the causes of the cycle. This is because all living things in an ecosystem are interconnected.

▶ CHECK YOUR UNDERSTANDING

1. How can competition affect the success of a plant or animal population?
2. Why do gardeners sometimes thin out rows of vegetable plants while the plants are still small?
3. Aboriginal peoples did not blame the lynx when the hare population declined. What might these Aboriginal peoples have known that the Hudson Bay Company fur traders did not?
4. Explain why a pet cat might have more effect than a wild predator on a population of birds.

1 Chapter Review

Ecosystems support life.

Key Idea: Ecosystems are made up of living and non-living things.

plants   animals   temperature   Sun   soil   water

Living Factors   Non-Living Factors

Vocabulary
organisms p. 6
micro-organisms
   p. 6

Key Idea: Groups of living things interact within ecosystems.

Vocabulary
species p. 6
population p. 6
community p. 6
ecosystem p. 7

Key Idea: All the ecosystems on Earth are interconnected.

Vocabulary
biosphere p. 12
biomes p. 12

Key Idea: Limiting factors determine which species' needs will be met in an ecosystem.

- Temperature
- Food
- Sunlight
- Shelter
- Water

Vocabulary
habitat p. 15

Key Idea: Living things interact in different ways.

Competition   Predator-Prey relationship

Vocabulary
predator p. 23
prey p. 23

Review Key Ideas and Vocabulary

When answering the questions, remember to use vocabulary from the chapter.

1. Make a two-column table with the headings "Living" and "Non-Living." Think about your local ecosystem. In your table, list as many living and non-living parts of your local ecosystem as possible.

2. Humans are a species. They are also a population in your local ecosystem. What are some other populations that form communities with humans in your local ecosystem?

3. A pond lies untouched by humans in a remote part of the province. Your local garden centre has a demonstration pond to show people how to create a water garden. How are these two ponds similar? How are they different?

4. Describe two ways in which your local land ecosystem is linked to the Pacific Ocean.

5. Infer two factors that might limit the size of a population of small plants growing on a forest floor.

6. Give an example of a predator-prey relationship from your local ecosystem. Is there another predator that competes for the same prey? If so, what is it?

Use What You've Learned

7. Go for a walk near your home or school and find a small ecosystem. Sketch the ecosystem, or take a photo of it. Make a presentation to your class, in which you describe the living and non-living parts of the ecosystem. Compare ecosystems with your classmates.

8. Identify a predator in your local ecosystem. Find out all the organisms it uses as prey. Is it prey for any other organisms? Use print or electronic sources of information, or ask knowledgeable people in your community.

www·science·nelson·com **GO**

9. Make a display about the ecosystem where you live. Your display could include
   - photographs of the landscape
   - samples, sketches, or models of typical plants and animals
   - samples of soils and rocks
   - graphs, models, or newspaper articles that describe the climate
   - newspaper articles or Aboriginal stories about how the environment affects the people who live in the ecosystem

Think Critically

10. Do you compete with any of the organisms in your environment? If so, how?

11. Are you a predator? Explain.

12. Do you think that the environment affects humans less, more, or about the same as it affects other organisms?

13. Humans can live in more types of environments than any other species. Explain why.

Reflect on Your Learning

14. Make a list of new things that you have noticed or learned about your local ecosystem.

15. List two questions that you still have about ecosystems. Glance through the rest of the unit. Do you think your questions will be answered in the topics that are covered? If not, where can you go to find the answers?

CHAPTER 2

Energy flows and matter cycles in ecosystems.

▶ The Sun is the source of all the energy in most ecosystems.

▶ Energy flows through ecosystems.

▶ A model can be used to show how energy flows through an ecosystem.

▶ A model can be used to show how the amount of energy that is available to organisms decreases at each level in a food chain.

▶ Matter cycles within ecosystems.

This orca needs a lot of materials to maintain its large body. Where do the materials that make up its body come from? Where do they go if the orca dies of disease or old age? Such a large animal also needs a great deal of energy for movement. Where does the energy come from? Where does the energy go when the orca dies? How does an ecosystem meet the needs of its many plants and animals? Why do ecosystems not run out of materials and energy?

In this chapter, you will study the flow of energy and the cycling of matter in ecosystems. Some of the things you will study, however, are difficult to observe in nature. As well, it is important not to disturb ecosystems. For these reasons, you will use models to track energy and materials through ecosystems.

All living things need food. An organism's role in an ecosystem depends on how it obtains its food because this affects how it interacts with other organisms in the ecosystem. The combination of where an organism lives (its habitat), how it obtains its food, and how it interacts with other organisms is called its **niche** [NEESH]. Plants and animals obtain food from their surroundings in two very different ways. Therefore, they have very different roles in an ecosystem.

Producers

Green plants make their own food using materials from the non-living environment. Light energy from the Sun reaches Earth, and plant leaves absorb this energy from the Sun. Along with this energy, plants use water from the soil and carbon dioxide from the air to produce their own food. This process is called **photosynthesis** [foh-toh-SIN-thuh-sis] (**Figure 1**). The food that plants produce is sugar and starches. Plants also produce oxygen, which is released into the environment. Humans and other animals breathe the oxygen that is produced by plants.

LEARNING TIP ◁

The process of photosynthesis is described in three ways on this page—in words, in an illustration, and in a chemical equation. Check to make sure you understand how each way matches the others.

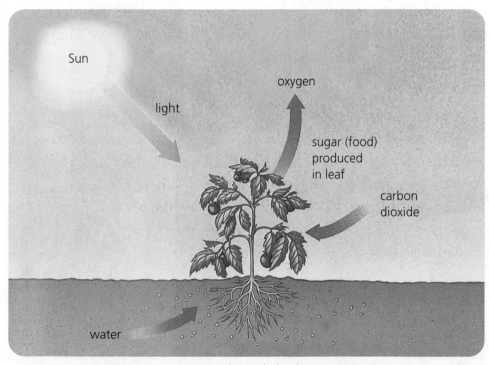

$$\text{carbon dioxide} + \text{water} \xrightarrow{\text{photosynthesis}} \text{food (sugar)} + \text{oxygen}$$

Figure 1
Plants make food through the process of photosynthesis.

Organisms that can make their own food from non-living materials are called **producers** (**Figure 2**). Producers include plants on land and in water.

Figure 2
This tree and this kelp are both producers.

Consumers

Animals cannot carry out photosynthesis. They must get their food from the living environment by eating, or consuming, other organisms. This is why animals are called **consumers.**

Consumers that eat plants are called **herbivores** (**Figure 3(a)** and **(b)**). Consumers that eat other animals are called **carnivores** (**Figure 4(a)** and **(b)**). Consumers that eat both plants and animals are called **omnivores** (**Figure 5(a)** and **(b)**).

Figure 3(a)
Deer are herbivores that feed on producers, such as grass and other plants, on land.

Figure 3(b)
Sea urchins are marine herbivores that feed on marine producers called kelp.

Figure 4(a)
A wolf is an example of a carnivore in a land ecosystem.

Figure 4(b)
An orca is an example of a carnivore in an ocean ecosystem. Why do you think orcas are sometimes called the "wolves of the sea" in Aboriginal legends?

Figure 5(a)
British Columbia's provincial bird, the Steller's jay, is an omnivore that eats insects, eggs, nuts, and seeds.

Figure 5(b)
The bat (or webbed) star is a marine omnivore that eats other sea stars, as well as worms and algae.

Detrivores and Decomposers

Not all plants and animals die because they are eaten. Some just die when their life span is over. Dead plants and animals become food. Organisms that feed on large bits of dead and decaying plant and animal matter are called **detrivores** (**Figure 6**). Crabs and some sea birds are the detrivores in ocean ecosystems. Earthworms, dung beetles, and wolverines are three examples of detrivores in land ecosystems.

Figure 6
Earthworms are common detrivores in land ecosystems, and crabs are common detrivores in ocean ecosystems.

Even detrivores, however, leave behind some waste materials: parts of the dead plant and animal matter and their own waste. Bacteria and fungi break down these waste materials. Organisms that get their food energy by breaking down the final remains of living things are called **decomposers.** Fruit rotting on the ground, a sandwich moulding in the bottom of a locker, and a shrinking pile of seaweed on the beach are all examples of decomposers at work (**Figure 7**).

Figure 7
Decomposers at work in an ecosystem

▐▶ CHECK YOUR *UNDERSTANDING*

1. Explain how producers and decomposers link the living and non-living parts of ecosystems.

2. In this section, you learned about six categories of organisms: producers, herbivores, carnivores, omnivores, detrivores, and decomposers. Describe how the organisms in each category obtain food from their environment.

3. In your notebook, make a chart like the one below. List examples of organisms in each category that live on land and in the oceans.

3.	Land	Oceans
producers	tree	
herbivores		
carnivores		
omnivores		Bat Star
detrivores		
decomposers		

4. What type of consumer are you? Are all people the same type of consumer?

Food Chains and Food Webs

Energy flows through ecosystems. When a herbivore eats a plant, the food energy that is stored in the plant passes into the herbivore's body. When the herbivore is eaten by another consumer, the food energy that is stored in the herbivore's body passes into that consumer's body. A model that shows how food energy passes from one organism to another in a feeding pathway is called a **food chain** (**Figure 1**). Each organism in a food chain depends on the organism before it in the chain for its food energy.

LEARNING TIP

Flow charts like the ones in **Figure 1** are used to show a sequence of steps or a timeline. Where else have you seen flow charts used?

Figure 1
The flow of energy through two different food chains: The arrows show the direction of the energy flow. What is the original source of energy for each food chain?

2.2 Food Chains and Food Webs

Consumers do not usually rely on only one source of food. For example, a coyote eats rabbits, but it will also eat mice, grouse and their eggs, and many other animals. The mice that the coyote eats consume the seeds, inner bark, and shoots of many different plants. Thus, most organisms are part of several food chains. A model that shows several different food chains, and the connections between them, is called a **food web** (**Figures 2** and **3**).

▷ **LEARNING TIP**

Evaluate the food web in **Figure 2** by comparing it to what you have learned about the Khutzeymateen Valley. Do you think that all of the organisms in the valley have been shown?

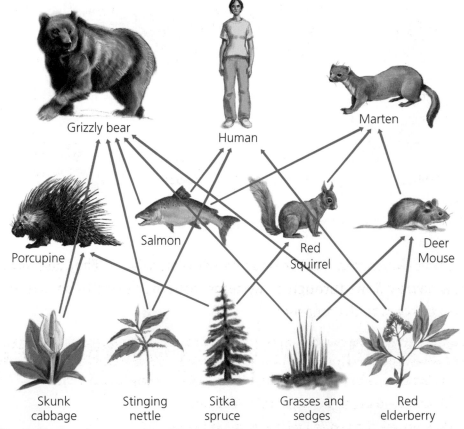

Figure 2
This food web shows some of the organisms in the Khutzeymateen Valley ecosystem. It is made up of many food chains.

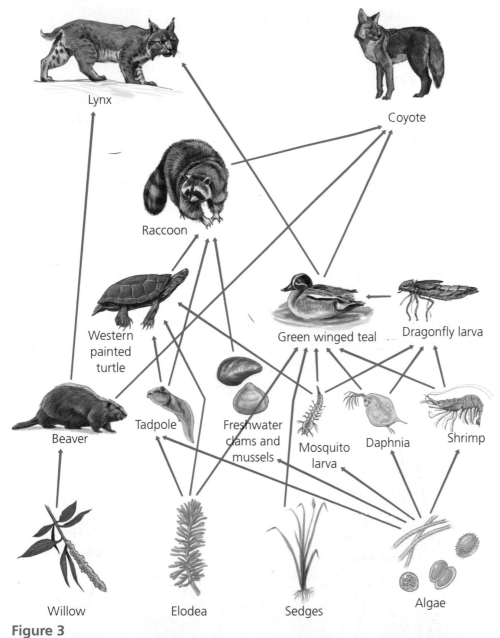

Figure 3
This food web shows some of the organisms in a pond.

Food chains and food webs show who eats whom. They also show how energy flows through ecosystems from producers to consumers to detrivores to decomposers.

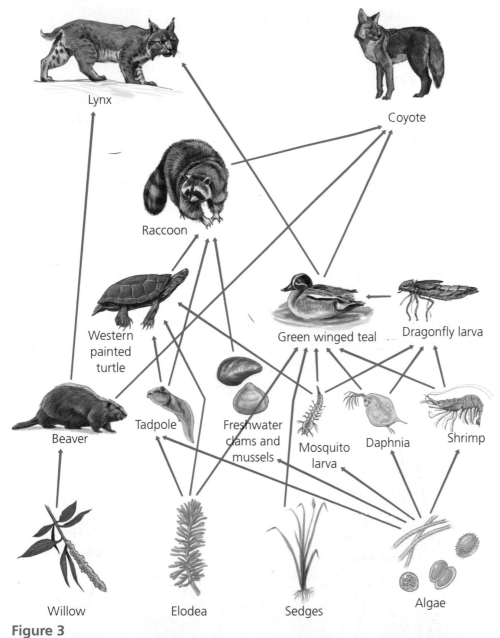

▶ CHECK YOUR UNDERSTANDING

1. What type of living thing does a food chain always begin with? Why?
2. Draw a food chain that ends with a pet.
3. How are food chains related to a food web?

SKILLS MENU

- ○ Questioning
- ○ Observing
- ○ Predicting
- ○ Measuring
- ○ Hypothesizing
- ● Classifying
- ○ Designing Experiments
- ● Inferring
- ○ Controlling Variables
- ● Interpreting Data
- ● Creating Models
- ○ Communicating

Modelling a Food Web

In this investigation, you will link a variety of different organisms in a food web.

Question

What are the links between producers, consumers, detrivores, and decomposers?

Materials

- square or rectangular piece of corrugated cardboard, at least 30 cm by 30 cm
- pen
- paper
- pushpins
- thread or string
- scissors

 Be careful with sharp objects such as pushpins and scissors.

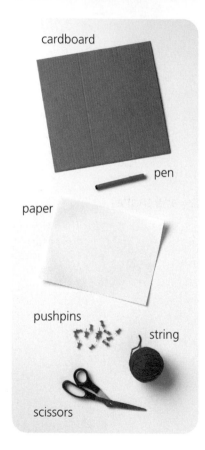

cardboard

pen

paper

pushpins

string

scissors

⏵ Procedure

1 Divide and label your cardboard as shown below.

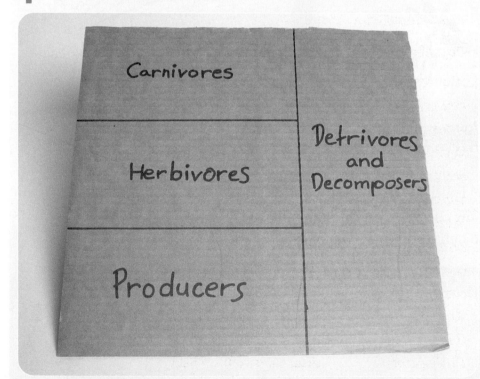

2 Choose one of the following three ecosystems:

Ecosystem in a Decaying Log

Organism	What it needs for food
decaying fallen tree	none now; used sunlight to make its own food when alive
mosses	sunlight to make their own food
ferns	sunlight to make their own food
beetles	decaying wood
fungi and bacteria	decaying wood, dead plants and animals, and animal waste
truffles	decaying wood
termites	decaying wood
mites	decaying wood
carpenter ants	decaying wood
spiders	termites, mites, and carpenter ants
newts	termites, mites, carpenter ants, and spiders
winter wrens	termites, mites, carpenter ants, and spiders
martens	voles, chipmunks, and birds' eggs
voles	moss, termites, carpenter ants, spiders, and truffles
chipmunks	young leaves and shoots of ferns; termites, carpenter ants, and truffles
owls	winter wrens, voles, and chipmunks

West Coast Ocean Ecosystem

Organism	What it needs for food
microscopic plants	sunlight to make their own food
kelp	sunlight to make its own food
microscopic animals	microscopic plants
sea urchins	kelp
herring	microscopic animals
salmon	herring
sea otters	sea urchins
seals	herring and salmon
humans	herring, sea urchins, salmon, crabs, and kelp
orcas	seals, sea otters, and sea birds
sea birds	herring and dead animals
crabs	dead animals
decomposers	dead plants and animals

Antelope Brush Ecosystem in the South Okanagan

Organism	What it needs for food
antelope brush	sunlight to make its own food
needle and thread grass	sunlight to make its own food
Behr's hairstreak butterflies	antelope brush for larvae (caterpillars)
mule deer	grass seeds and antelope brush leaves and twigs
Great Basin pocket mice	grasses and seeds from antelope brush
burrowing owls	Great Basin pocket mice, scarab beetles, and grasshoppers
Northern Pacific rattlesnakes	Great Basin pocket mice and ground nesting birds (western meadowlarks)
scarab beetles	animal waste and decaying plants and animals
western meadowlarks	beetles, caterpillars, and grasshoppers
prairie falcons	burrowing owls and western meadowlarks
grasshoppers	grasses and leafy plants
humans	mule deer
cougars	mule deer

3 Decide whether each organism is a producer, herbivore, carnivore, detrivore, or decomposer. Put a pushpin in the section where your organism belongs. If it fits in two sections, for example, if it is an omnivore, use two pushpins. Label the pushpin by writing the name of the organism on the cardboard beside it or by attaching a "flag" to the pushpin.

4 Use the thread or string to connect each organism to the organisms it feeds on. Wrap the thread or string around the pushpin a couple of times so that it does not fall off.

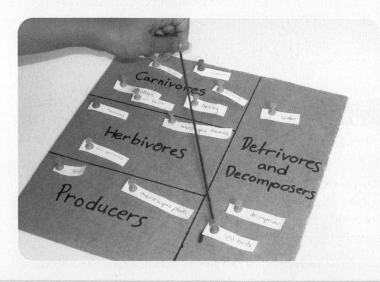

Analyze and Evaluate

1. Your food chains all begin with producers. Where do the producers obtain their food?

2. Describe the effect on other organisms in your food web if you remove
   - one of the producers
   - one of the herbivores
   - one of the carnivores

Apply and Extend

3. A farmer or gardener uses a toxic spray against insect pests. A few weeks later, a large quantity of the insecticide is found inside the body of a dead hawk that hunted in the area (**Figure 1**). Explain, using a diagram, how the insecticide might have found its way into the hawk's body.

Figure 1
The number of Swainson's hawks is declining because of the use of pesticides in Argentina, where it winters.

▶ CHECK YOUR UNDERSTANDING

1. Models are used to explain scientific concepts. What concept did your food web explain?

2. Models provide an opportunity to show what may happen in a situation that you cannot easily observe. How did your food web provide an opportunity for you to observe something that you would not normally have an opportunity to observe?

3. Any model has limitations. What limitations did your food web have?

LEARNING TIP ◁

For further information on creating and using models, see "Creating Models" in the Skills Handbook.

Hydrothermal Vents

In 1977, scientists were exploring the ocean in a submersible called *Alvin* when they found a deep-sea hot spring, or hydrothermal vent.

The vent was 2.5 km below the surface of the ocean. The scientists were not surprised to find the vent since they had predicted that vents existed. They were very surprised, however, to find it surrounded by large numbers of strange animals, most of which had never been seen before. Since then, hydrothermal vents have been discovered in many parts of the ocean, including west of Vancouver Island, British Columbia.

Sunlight cannot reach the deep parts of the ocean, where hydrothermal vents have been discovered. Therefore, no plants can photosynthesize. But if there are no plants, what are the producers in these deep-sea ecosystems?

Scientists have discovered that the producers at hydrothermal vents are bacteria. These bacteria can make food from chemicals that are released at the vents. The process of making food from chemicals is called chemosynthesis. Animals, such as limpets and mussels, consume the chemosynthetic bacteria. Predators, such as octopuses and vent crabs, prey on the limpets and mussels. The vent crabs also serve as detrivores, making a complete food web based on the chemosynthetic bacteria.

The most fascinating creatures that are found at the vents are tubeworms (**Figure 1**). Adult tubeworms have no mouths or stomachs. They survive because the chemosynthetic bacteria live inside them.

Figure 1

The ecosystem at this hydrothermal vent is based on producers that use chemicals, rather than light, to make food.

As energy flows through ecosystems, from producers to consumers to detrivores to decomposers, some energy is lost at each level.

The Sun is life's main energy supply. Using energy from the Sun, plants make their own food through the process of photosynthesis. Plants need to use most of the energy from the food they make for everyday life processes, such as growing and producing flowers and seeds. On average, only about one-tenth ($\frac{1}{10}$ or 10%) of a plant's food energy gets stored as nutrients in the roots, leaves, and other parts of the plant. So, when a plant is eaten by a consumer, such as a deer, only one-tenth of its energy is available to the consumer.

Similarly, the deer uses most of the energy from its food (the plant) to support its everyday life functions, such as breathing, moving, and chewing. Energy is also is given off as body heat. Consequently, when the deer is eaten by a consumer, such as a cougar, only about one-tenth of its energy is available to the consumer. Thus, very little energy is passed on from one organism to the next in a food chain (**Figure 1**).

LEARNING TIP ◁

Try to make a mental picture of how energy enters and leaves your own body.

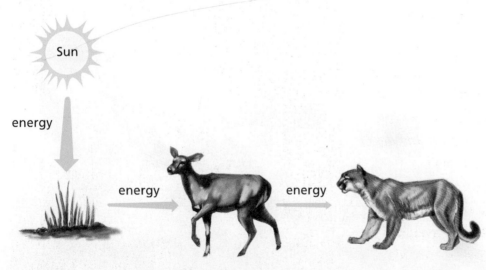

Figure 1
Energy flow through an ecosystem

TRY THIS: *MODEL ENERGY LOSS*

Skills Focus: creating models

1. Form groups of three and assign the following roles: producer, herbivore, and carnivore.

2. The producer takes ten sheets of paper from the recycling box and spreads them out in a row on the table. This represents the amount of energy from the Sun that the producer has stored as food.

3. The herbivore takes one-tenth of the producer's energy (one piece of paper) from the producer and puts it on the table above the producer's papers.

4. The carnivore takes one-tenth of the herbivore's energy by tearing off one-tenth (a 2-cm strip) of the herbivore's paper and putting it on the table above the herbivore's paper.

5. As a group, calculate the percentage of the energy in the producer that was transferred to
   a) the herbivore
   b) the carnivore

The model you made to show energy loss in a food chain is called an **ecological pyramid** (**Figure 2**).

LEARNING TIP

As you study **Figure 2**, ask yourself, "What is the purpose of this model? What do scientists use it to illustrate? What am I supposed to notice and remember?"

Figure 2
The base of the pyramid holds producers (plants). At each level above the producers, the amount of available energy is reduced. This explains why, in an ecosystem, you might find a huge number of insects to eat the plants, a much smaller number of shrews to eat the insects, and only a very few owls to eat the shrews.

Each level of an ecological pyramid matches a level of producers or consumers in a food chain. At each level, the amount of available energy is less than the amount of available energy in the level below. Usually the number of organisms also decreases at each higher level of the pyramid (**Figure 3**).

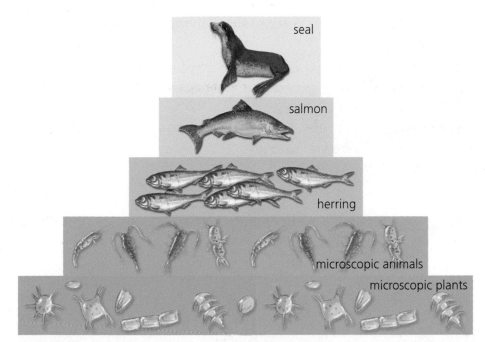

Figure 3
An ecological pyramid in an ocean ecosystem.

LEARNING TIP ◁

Check your understanding of ecological pyramids by comparing **Figures 2** and **3**. How are they the same? How are they different?

⫸ CHECK YOUR UNDERSTANDING

1. Why is some energy lost at each level in a food chain?
2. Using your own words, describe why there are usually fewer large carnivores than herbivores or producers in an ecosystem?

2.5 The Role of Decomposers in Recycling Matter

▷ **LEARNING TIP**

Make connections to your prior knowledge. What do you already know about decomposers?

Earth is often compared to a spaceship (**Figure 1**). It has been launched, and nothing more can be added to it. Because a spaceship is closed, air and water must be recycled or the astronauts will die. The same is true on Earth. Life depends on the recycling of matter. How does this recycling occur? How much do we depend on the recycling of matter? What is the role of living things in the recycling of matter?

Figure 1
Life on this spaceship depends on careful recycling of matter, as does life on Earth.

Food chains and food webs show how matter and energy are moved from one organism to another. We often forget, however, about a very important part of this cycle: the decomposers (**Figure 2**). As decomposers break down their food, they use the last of the energy in the food chain. They also release nutrients. Nutrients are chemical substances that organisms need to grow and survive. Nutrients are released into the soil, water, or air. They can be taken up by plants and used again to help the plants grow. Decomposers keep matter moving between the living and non-living parts of an ecosystem.

Figure 2
These bracket fungi are decomposers.

The importance of decomposers in an ecosystem should not be underestimated based on their small size. Imagine what your schoolyard would look like with years of accumulated leaves and grass clippings still in their original forms. Without decomposers, nutrients would remain locked in the tissues of dead plants and animals. Decomposers break down matter and turn it into the nutrients that living things need every day (**Figure 3**).

Figure 3
Moulds and bacteria spoil food, but by doing so they recycle nutrients within the ecosystem.

Composting

If you compost your kitchen scraps or plant waste, you are relying on the work of decomposers to break down the waste and release the nutrients it contains. In a compost ecosystem (**Figure 4**), small detrivores, such as earthworms, mites, grubs, insects, and nematodes (microscopic worms), chew, digest, and mix the waste. As detrivores mix the waste, air is added to the compost mixture. Decomposers, such as bacteria and fungi (moulds), then help to break down the waste further. This makes the nutrients available to plants when you put the compost on a garden. Putting compost on a garden is like giving the soil a giant vitamin pill.

Figure 4
A food web of a compost heap

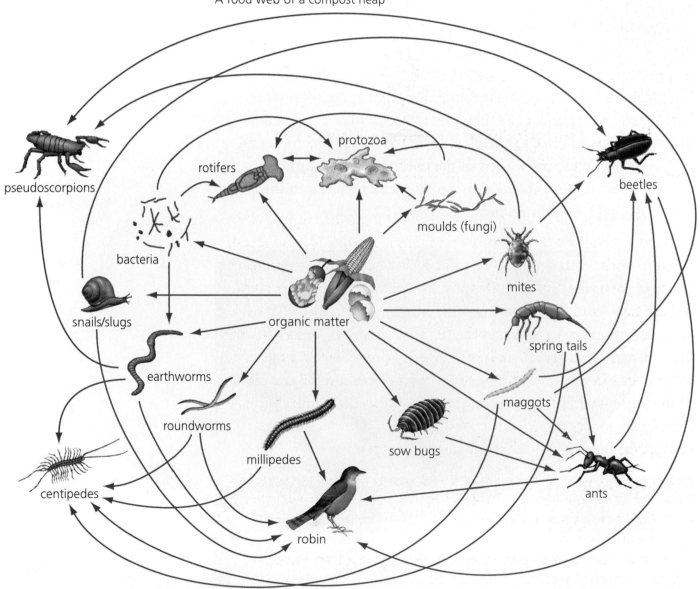

Dying Salmon

Salmon return from the ocean to their home stream to reproduce and die (**Figure 5**). Some onlookers are sad to see the masses of dying fish. The death of the salmon, however, helps to ensure the survival of the species. First detrivores (such as gulls, eagles, and bears) take their share of the dying and dead salmon. Then bacteria and fungi finish decomposing the salmon and turn the tissue into a liquid. This allows the nutrients from the salmon to be dissolved in the stream. In the spring, the nutrients in the stream help to nourish plankton, an important part of the salmon fry's food chain. If you visit a salmon stream in the spring, you will no longer find piles of rotting salmon, thanks, in part, to these decomposers.

LEARNING TIP

Close your eyes and try to "see" the process of the nutrients from the salmon being returned to the ecosystem.

Figure 5
Spawning sockeye salmon in the Adams River

The nutrients from the decomposing salmon are not only used by the next generation of salmon. The nutrients also fertilize the forest. Scientists believe that some forests have high levels of nutrients in the soil because of dead salmon, deposited there by feeding bears and wolves. Bears and wolves don't eat the whole salmon. The rest is left to be consumed by crows and other detrivores, and by decomposers. The nutrients that are released by the decomposers fertilize trees and other plants in the forest. Many of these nutrients entered the bodies of the salmon while they lived in the open ocean. The migration of the salmon moves these nutrients from ocean to forest, linking ecosystems that are thousands of kilometres apart.

⫸ CHECK YOUR UNDERSTANDING

1. Explain how decomposers link the living and non-living parts of an ecosystem.
2. What would happen if there were no decomposers in an ecosystem?

▶ SKILLS MENU

○ Questioning   ● Observing
○ Predicting   ○ Measuring
○ Hypothesizing   ○ Classifying
○ Designing Experiments   ● Inferring
○ Controlling Variables   ● Interpreting Data
○ Creating Models   ● Communicating

apron

pop bottle

vegetable and fruit scraps

garden waste

soil

water in spray bottle

paper

rubber band

tape

black marker

stick

Composting

Decomposers are a very important part of any ecosystem. They reduce matter and make nutrients available to the soil. In this investigation, you will observe decomposers at work on dead plant materials.

Question

How long does it take decomposers to break down dead plant materials?

Materials

- apron or old oversized T-shirt
- bottom half of clear plastic pop bottle
- garden waste (such as dead leaves, grass, pine needles, or straw)
- raw vegetable and fruit peels and scraps, cut into small pieces
- soil
- water in a spray bottle
- paper or screening
- rubber band
- tape
- black marker
- stick for stirring

 Some students have serious allergies to some decomposers. Minimize the time your compost is uncovered. Do not breathe the air over your set up.

Procedure

1 Put on your apron or old T-shirt.

2 In the bottom of the plastic bottle, put a 5-cm layer of garden waste, then a 5-cm layer of fruit and vegetable scraps, then a 5-cm layer of soil. Add more layers in the same order until the bottle is almost full (**Figure 1**). Decomposers work more quickly when they have a variety of "food." Dry and brown materials, such as dead leaves and pine needles, provide some nutrients. Green and wet materials, such as vegetable scraps, supply other nutrients. Do not add meat, bones, grains, pasta, dairy products, or fatty foods to your compost. If you do, it will get very smelly and may attract rodents.

3 Add enough water to make the mixture damp but not wet. To keep out flies, make a cover from paper or screening, and a rubber band. Using the tape and black marker, label the bottle with your name. Put the bottle in a warm place for one week.

4 In your notebook, make a table like the one below.

Figure 1
Composting in a plastic bottle

soil
fruit and vegetable scraps
garden waste

5 After a week, stir the mixture in the bottle and examine the materials. In your notebook, record how the materials look and smell. (Remember not to put your nose directly over the bottle.) If a camera is available, take a photo to help you accurately record the changes over time.

6 Make sure that the mixture is still damp. If necessary, add a small amount of water. Put the bottle in the warm place again. Wash your hands well.

7 Repeat steps 4 to 6 each week, until you can see changes occurring.

8 Describe the materials in the bottle after four weeks.

9 When you have finished your investigation, ask your teacher where you should add your compost to the soil.

Analyze and Evaluate

1. How long did it take for the dead plant materials to start decomposing? How long did it take for them to seem completely decomposed?

2. You make many inferences every day. For example, when you see your physical education teacher getting out soccer balls, you infer that you will be learning ball skills. Make an inference about this composting investigation, using the following questions to help you:
   • Did you put decomposers in the bottle?
   • How do you think the decomposers got in the bottle?

3. What kinds of decomposers do you think were in your bottle?

Apply and Extend

4. What are the advantages of backyard composters, for both homeowners and the community (**Figure 2**)?

Figure 2
Decomposers are turning kitchen waste into soil in this backyard composter.

⫸ CHECK YOUR UNDERSTANDING

1. Why is it important to read and follow the procedure in an investigation carefully?

2. During this investigation, you made and recorded observations weekly, using your senses. Explain why it is important to include qualitative data, such as your observations, rather than just stating that changes have or have not occurred.

Dollars from Decomposers

Wastewater is usually treated with chemicals so we can use it again. In this section you will learn how scientists are treating wastewater naturally, and making money too.

It's true! There is money in sewage! Tiny organisms at one of the wastewater treatment plants at the University of British Columbia (UBC) are making money every day. The treatment plant uses naturally occurring bacteria, instead of chemicals, to decompose waste and clean the wastewater. This means that money is not needed to buy chemicals for treatment, nor to dispose of chemical sludge at the end of the process. The bacteria also help to make a saleable product. By controlling certain environmental conditions, such as available oxygen, the bacteria will remove specific particles, such as nitrogen and phosphorus, from the sewage. The removed nitrogen and phosphorus can be processed to make fertilizers, such as those that are sold in garden centres. The UBC treatment plant can thus make money by selling these fertilizers (**Figure 1**).

Many cities and towns in the Prairie provinces are using this method to treat their wastewater. As well, Kelowna, Summerland, and Salmon Arm in British Columbia are using this method. Most of the treatment plants are not yet set up, however, for processing the nutrient sludge to make fertilizers. The communities are using this method because it is less costly for both taxpayers and the environment.

Figure 1
Professor Don Mavinic (left) and Fred Koch (right) with phosphorus fertilizer made from nutrient sludge.

The Water and Carbon Cycles

Plants constantly trap new energy from the Sun. This new energy replaces the energy that was lost as heat in every level of the food chain. In this way, energy is constantly being added to ecosystems. Although new energy continues to arrive from the Sun, no new water arrives. The water that is on Earth now is the same water that was here when the dinosaurs lived. It has been reused and recycled many, many times.

The Water Cycle

Anything that happens over and over again, like the seasons of the year or the phases of the moon, is called a cycle. Water moves through a **cycle** (**Figure 1**).

> ▷ **LEARNING TIP**
>
> Follow the arrows on the diagram as you read about the water cycle. Check your understanding by using the diagram to explain the water cycle to a partner.

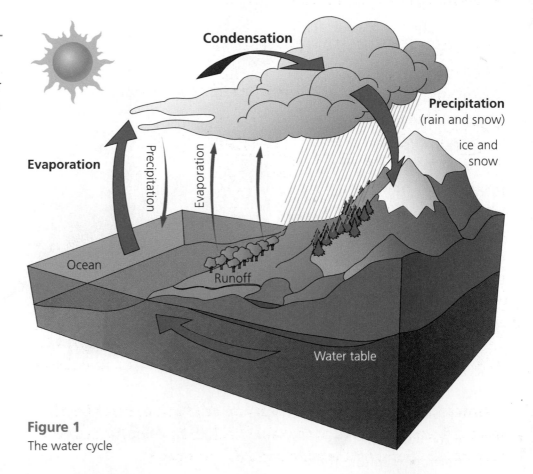

Figure 1
The water cycle

Energy from the Sun warms Earth's surface and causes water to evaporate from oceans and lakes. Water turns into water vapour when it evaporates, and the water vapour enters the atmosphere. In the atmosphere, the water vapour cools and changes back to liquid water in the form of clouds (condensation). The water then returns to the surface of Earth as rain or snow (precipitation). Some rain and melting snow sinks into the ground. This groundwater seeps down through the rocks and soil to the water table. It may remain underground for many years, but eventually returns to the ocean. Some rain and melting snow runs off into rivers. The rivers flow into lakes and oceans, where the cycle begins again.

Water cycles through the living parts of ecosystems as well as through the non-living parts. Animals drink water, which later leaves their bodies as urine or sweat. Plants take up water from the soil with their roots. Much of the evaporation of water in land ecosystems occurs from the leaves of plants (**Figure 2**).

LEARNING TIP ◁

Work with a partner. Read aloud the two paragraphs about the water cycle while your partner follows the explanation on **Figure 1** with his or her finger. Then, switch roles.

Figure 2
The many leaves in this British Columbia coastal rain forest return a large amount of water to the atmosphere.

How much water goes from plants into the atmosphere? Do all types of leaves lose water at the same rate? To find out the answers to these questions, try the activity on the next page.

Water does not stay in the atmosphere or the bodies of plants and animals very long. On the other hand, water can spend thousands of years in the deepest parts of oceans and trapped in glaciers.

Skills Focus: observing, measuring, hypothesizing

LEARNING TIP

For help with hypothesizing, see the Skills Handbook section "Hypothesizing."

Find a tree with large leaves. Put a small plastic bag around one of the leaves (**Figure 3**). Use a twist tie to gently close the bag. Put another bag over a small branch of a tree with needles (**Figure 4**). Leave the bags on the trees overnight. The next day, carefully pour the water from each bag into a graduated cylinder and measure it. Which type of leaf lost the most water? Create a hypothesis to explain the difference.

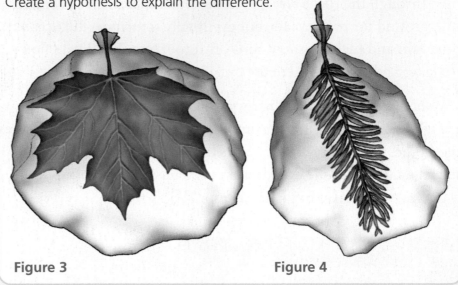

Figure 3   **Figure 4**

The Carbon Cycle

Just as no new water arrives on Earth, no new carbon arrives. Carbon is found in many parts of our world. For example, it is found in the chemicals that make up rocks. It is found underground in coal, oil, and natural gas. It is also found in the air in the form of carbon dioxide. Plants use carbon dioxide to photosynthesize and produce food. When animals break down this food to produce energy, they produce water and carbon dioxide as well, which plants can then use. All living things contain carbon. Decomposers that break down dead plants and animals return the carbon to the non-living parts of ecosystems. The carbon cycle (**Figure 5**) illustrates how carbon moves throughout ecosystems.

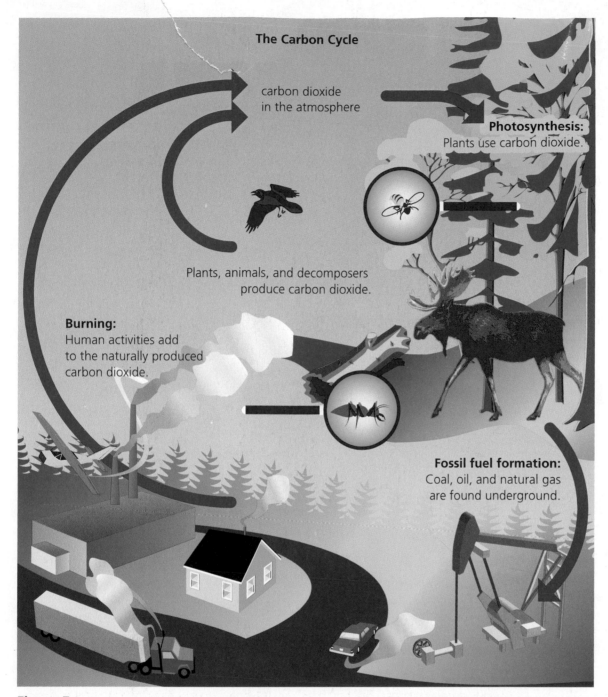

The Carbon Cycle

carbon dioxide
in the atmosphere

Photosynthesis:
Plants use carbon dioxide.

Plants, animals, and decomposers
produce carbon dioxide.

Burning:
Human activities add
to the naturally produced
carbon dioxide.

Fossil fuel formation:
Coal, oil, and natural gas
are found underground.

Figure 5
The carbon cycle

1. If no new water ever arrives on Earth, where do rain and snow come from?

2. Give three examples of sources of carbon in your schoolyard.

3. Explain what would happen to the carbon cycle if there were no decomposers?

Chapter Review

Energy flows and matter cycles in ecosystems.

Key Idea: The Sun is the source of all the energy in most ecosystems.

Vocabulary

niche p. 29

photosynthesis p. 29

producers p. 30

Key Idea: Energy flows through ecosystems.

Vocabulary

consumers p. 30

herbivores p. 30

carnivores p. 30

omnivores p. 30

detrivores p. 31

decomposers p. 32

Key Idea: A model can be used to show how energy flows through an ecosystem.

Vocabulary

food chain p. 33

food web p. 34

Key Idea: A model can be used to show how the amount of energy that is available to organisms decreases at each level in a food chain.

Vocabulary

ecological pyramid p. 42

Key Idea: Matter cycles within ecosystems.

Vocabulary

cycle p. 52

Review Key Ideas and Vocabulary

When answering the questions, remember to use vocabulary from the chapter.

1. How does energy enter the food chain?

2. Your friend tells you that all the energy you get from eating a pizza comes from the Sun. Draw a food chain to explain this statement.

3. The gopher snake lives in southern British Columbia. It eats small rodents and sometimes eggs and young birds from birds' nests. Rodents eat seeds and plants, including crops that humans grow to feed the cattle they raise for food. Hawks eat snakes and birds. Draw a food web using all of these organisms.

4. Explain how energy is lost at each level in a food chain.

5. What role do plants play in the water cycle? What role do they play in the carbon cycle?

Use What You've Learned

6. Wolves do not eat plants, but they could not live in an ecosystem that did not have plants. Explain.

7. Draw two food chains that you could find in your local ecosystem—one that is land based and another that is water based. Draw an ecological pyramid for each food chain.

8. Aboriginal people recognize the importance of healthy eel-grass beds in the coastal ecosystem. The Nuu-Chal-Nulth people are working with scientists to re-establish eel-grass beds. Research the living and non-living parts of an eel-grass bed

ecosystem. Then draw a possible food web for this ecosystem.

www.science.nelson.com GO

9. In a small group, research what producers, herbivores, carnivores, omnivores, decomposers, and detrivores are found in your local ecosystem. Record your results in a table like those in section 2.3. Then use your table to make a food web for your local ecosystem.

10. Contact your nearest municipality and ask for a brochure on the wastewater treatment plant. Does the plant use decomposers? If not, how does the waste get broken down?

11. Design a board game to teach people about the interactions within a food web.

Think Critically

12. Explain the meaning of the following statement: The importance of decomposers is out of proportion to their size.

13. A keystone is a stone at the top of an arch. It supports the other stones and keeps the entire arch from falling. A keystone species is a species in an ecosystem that many other species in the ecosystem depend on for survival. Look over the food webs for different ecosystems in this chapter. Which species do you think is the keystone species in each ecosystem? Explain why.

Reflect on Your Learning

14. Make a list of things you learned in this chapter. Put an asterisk (*) beside any that surprised you. Why did they surprise you?

CHAPTER 3

Human survival depends on sustainable ecosystems.

▶ **KEY IDEAS**

- ▶ Biodiversity keeps ecosystems strong and stable.

- ▶ Indigenous knowledge can help us achieve sustainability.

- ▶ Human activities can decrease biodiversity.

- ▶ Humans can preserve, conserve, and restore ecosystems.

- ▶ Humans can lessen their ecological footprint.

Ribbit! These rare tailed frogs were lucky to have someone to speak up for them. Some students in West Vancouver found tailed frogs in an area where developers planned to build houses. The students' interest encouraged the property owners to plan for the frogs' survival. The owners supported a five-year study to determine where tailed frogs existed in the area. This is an example of how humans can work together to plan for the future of an ecosystem.

Endangered spaces usually mean endangered species. Is there a species of plant or animal in your community that needs your voice? Is there a space—a stream, a field, or a schoolyard—in your community that needs your voice? You, like the students in West Vancouver, can be a voice for endangered spaces and species.

The Role of Humans in Ecosystems

What is your place in your ecosystem? Do you visit nature, or do you think of it as your home? Do you worry about your effect on other species, or do you think non-human species are less important than you? Should people control ecosystems, or should people control their use of ecosystems? These are very complicated questions. How people live in ecosystems depends on how they see themselves and their world (**Figure 1**).

The ecosystems of the natural world support the web of life. Energy flows from the Sun, and one species' waste is another species' food. Matter cycles through ecosystems, creating a balance. Like all other living things, humans depend on ecosystems to provide their needs for survival and to recycle wastes.

LEARNING TIP

Work with a partner and discuss the questions in the first paragraph. Compare your reactions to those of your partner.

Figure 1
Do you think you can "conquer" nature?

For most of human history, ecosystems were able to sustain, or bear the weight of, their human populations over thousands and thousands of years. This ability of ecosystems is called **sustainability.** A sustainable ecosystem is able to replace the resources that people remove, and recycle the wastes that people put in.

Humans are a very successful species. As our population continues to grow, so does our need for food, air, water, and shelter. Also, we produce more waste. Today, we often remove resources faster than ecosystems can replace them. We often put wastes into ecosystems faster than the natural cycles can deal with them. As a result, some ecosystems can no longer provide for other species. The other species die or leave.

Each time a species is lost from the web of life, the ecosystem is weakened. A strong ecosystem has a variety of species and complex interactions. The variety of species in an ecosystem is called **biodiversity.**

Scientists know that a loss of biodiversity can threaten the sustainability of our ecosystems. This is such an important problem in our world today that the United Nations has developed a Convention on Biodiversity. It asks all the countries of the world to pay more attention to the knowledge and ways of indigenous peoples, the peoples who originally lived in an area (**Figure 2**).

"...respect, preserve and maintain knowledge, innovations and practices of indigenous and local communities embodying traditional lifestyles relevant for the conservation and sustainable use of biological diversity and promote their wider application with the approval and involvement of the holders of such knowledge, innovations and practices and encourage the equitable sharing of the benefits arising from the utilization of such knowledge, innovations and practices..."

Figure 2
The UN Convention on Biodiversity recognizes the important role of indigenous knowledge.

The longer you observe something, the more you learn about it. Aboriginal peoples have observed local ecosystems for thousands of years. They have passed their knowledge from generation to generation, often in the form of stories (**Figure 3**). Stories are useful models for complex relationships, like those found in nature. Aboriginal stories emphasize the relationships among living things and their non-living environment. As a result, Aboriginal peoples often have a very good understanding of how interconnected all things are, and how important it is for humans to live in harmony with other living things and the environment. The United Nations has decided that this is the type of understanding we need to have to ensure the sustainability of our ecosystems.

LEARNING TIP ◁

Think of other examples in which people use stories to communicate scientific information.

Figure 3
Aboriginal Elders pass on their knowledge through stories.

If you listen to Aboriginal stories, you will notice that nature usually has something to teach people. The stories emphasize that nature is in control, not people. Scientists who study ecosystems have learned the same thing.

⏵ CHECK YOUR UNDERSTANDING

1. What do humans remove from ecosystems, and what do they put in? How is this the same as what other species do? How is this different?

2. Would you expect to find greater biodiversity in a natural ecosystem or a human-made ecosystem, such as a farm or a garden? Explain your answer.

3. How can indigenous knowledge help all British Columbians today?

3.2 Pollution of Ecosystems

▷ **LEARNING TIP**

Look at the headings in this section. What three types of pollution will you be studying? Make a three-column chart using these headings and record examples as you read.

A comedian once said that pollution is a dirty word. Pollution refers to things that are not normally found in the environment, or things that are present in such large amounts they overload the natural cycles in an ecosystem. Do you often get the feeling that others, such as manufacturers, farmers, and people driving to work, are polluting your environment (**Figure 1**)? We all, however, pollute the environment. Think of your food and home, and how you travel each day.

Pollutants in the land, air, or water can enter food webs and end up in the bodies of plants and animals.

Figure 1
Do you see evidence of pollution?

Polluting the Land

Most of the household garbage that you produce ends up in landfill sites (**Figure 2**). A landfill is created by digging a very large hole and, over a period of time, filling it with garbage. The original ecosystem of the field or forest is replaced by the ecosystem of the landfill site. If electronics waste (such as computers and batteries) and poisonous chemicals are put in a landfill, these harmful substances can pollute the soil and water. The environment changes and the plant and animal species that used to live in the ecosystem die or leave.

Figure 2
These seagulls are feeding at the Vancouver Landfill in Delta.

Landfill sites contain many decomposers. As well, they sometimes attract detrivores, such as bears, seagulls, and rats. There is usually less biodiversity, however, than in the original ecosystem. The more garbage people throw out, the more land that is used for landfill sites, and the more biodiversity that is lost. Composting and recycling are two ways to reduce the amount of waste that we put in landfills (**Figure 3**).

Figure 3
How can recycling and composting protect biodiversity in your ecosystem?

Polluting the Air

Have you looked at the labels on your food, clothing, and electronics? If these products were made far from your home, how did they travel? Most of them have likely been on a ship, plane, train, or transport truck and have travelled hundreds or even thousands of kilometres. The leading cause of air pollution is the transportation of goods and people.

Air pollution can cause health problems for humans, other animals, and even plants. It can also cause acid rain. Acid rain can harm the soil that plants need to survive. Acid rain can also increase the acidity in lakes. This reduces biodiversity, since most plants and animals cannot survive in acidic lakes.

Air pollution can even change the climate. A certain amount of carbon dioxide in the atmosphere is needed for photosynthesis. It helps to trap the Sun's heat energy, just like greenhouse glass traps the Sun's heat energy, warming the greenhouse. Scientists call this the greenhouse effect. Without the greenhouse effect, Earth would be too cold for life as we know it. The burning of coal, oil, and natural gas changes the balance in the carbon cycle so that there is more carbon dioxide in the atmosphere. Scientists hypothesize that the extra carbon dioxide is causing global warming. Even a slight increase in the average temperature can affect climates worldwide, changing the water levels in oceans and creating extreme weather patterns of heat and cold, and droughts and floods (**Figure 4**). What effects might these changes have on people, crops, and animals?

Figure 4
Sachs Harbour is located on Banks Island in Canada's Western Arctic. Warmer temperatures are melting the permafrost, causing the ground to sink and shift. This can be clearly seen at the edge of this lake near the coast of Banks Island.

Polluting the Water

When you flush your toilet or drain your sink, the wastewater has to go somewhere. It may go to a sewage treatment plant on its way to a river and eventually to the ocean. Every year, billions of litres (L) of untreated sewage from cities and communities pour into waterways and then into the oceans worldwide. About 80% of ocean pollution is caused by human activities on land. Agriculture and industry also contribute to water pollution during the processes that produce the food you eat and the products you buy. Water pollution kills many plants and animals, reducing the biodiversity in our lakes, rivers, and oceans (**Figure 5**).

Figure 5
How do you think the garbage in this photo is affecting the water?

⫸ CHECK YOUR UNDERSTANDING

1. Which types of pollution affect you personally? Explain how you have been affected by each.

2. Explain how your household garbage can reduce biodiversity.

3. Explain how your shopping habits could contribute to air pollution.

4. Estuaries are important ecosystems for migratory birds. Explain how pollution in your community could affect an estuary ecosystem hundreds of kilometres away.

5. Which types of pollution mentioned in this section do you think are occurring in your community?

SKILLS MENU

- ○ Questioning
- ● Observing
- ○ Predicting
- ● Measuring
- ● Hypothesizing
- ○ Classifying
- ○ Designing Experiments
- ○ Inferring
- ● Controlling Variables
- ● Interpreting Data
- ○ Creating Models
- ● Communicating

Polluted Waters

When streams and lakes are polluted, the organisms that live in the water are affected. Can you think of some ways that a plant or animal may be affected by a change in its watery home? Can some organisms still live in polluted water?

Earlier in this unit, you learned that yeast (a living thing) lives best in warm, sugary water. In this investigation, you will observe how yeast is affected when a pollutant is added to its non-living environment. The pollutant you will add is lemon juice, an acid.

Question

How will different amounts of a pollutant affect the growth of an organism that lives in water?

Hypothesis

Write a hypothesis that answers the question above. Use the following form for your hypothesis: "If … then …."

Materials

masking tape
beaker
apron
pen
measuring spoons
goggles
sugar
measuring cup
water
lemon juice
soup spoons
large jar
baker's yeast
watch

- 10 250-mL clear glass jars or beakers
- masking tape
- pen or marker
- apron
- safety goggles
- measuring spoons
- measuring cup
- 5 mL of sugar for each small container
- graduated cylinder
- very warm (not hot) water for each small container
- lemon juice at room temperature
- 2 soup spoons
- 1 large jar (500 mL)
- 100 mL of very warm (not hot) water for larger jar
- 25 mL of instant rise baker's yeast
- clock or watch

 Wear safety goggles for this investigation.

Procedure

1 In your notebook, make a table like the one below.

Effects of Lemon Juice on Yeast Growth							
	3 min	5 min	7 min	9 min	11 min	13 min	15 min
none							
5 mL							

2 Working with a partner, set up your containers in a line. Label the first container "none." Then, label the rest by fives: 5 mL, 10 mL, 15 mL, and so on until you have labelled all of them.

3 Put 5 mL of sugar in each container. Using a graduated cylinder, add 50 mL of very warm water to the first container, 45 mL to the second, and 40 mL to the third. Continue to decrease the amount of water added by 5 mL until the tenth container has only 5 mL of water added to it. Then, put lemon juice in each container, as indicated on the label. In other words, put no lemon juice in the container labelled "none," put 5 mL in the container labelled "5 mL," and so on. Using a soup spoon, stir each container to dissolve the sugar. Work quickly but carefully.

4 In the large jar, put 100 mL of very warm water. Sprinkle the 25 mL of yeast into this jar, a little at a time. Stir with the other soup spoon after each addition.

5 Add 10 mL of this yeast mixture to each container. Swirl each container to mix. Note the time on the watch or clock.

6 Make your first observation 3 min after you add the yeast mixture to the containers. Continue to make observations every 2 min, until 15 min have passed. Record your observations in your table.

Analyze

1. Each container has a different amount of lemon juice. Was there a difference between any two containers that you think was important?

2. What did you learn from this investigation about the amount of a pollutant in water and the effect on the organisms that live in the water?

Write a Conclusion

3. Write a conclusion that explains the results of your investigation. Your conclusion should refer back to your hypothesis. Was your hypothesis correct, partly correct, or incorrect? Explain how you arrived at your conclusion.

Apply and Extend

4. Examine the photos in **Figure 1**. Which lake do you think has more acid in the water?

Figure 1

5. What do you know about acid rain? Write a brief paragraph about what it is and how it is currently affecting eastern Canada. You may need to do some research in a library or on the Internet.

www·science·nelson·com

▷ **LEARNING TIP**

For more information on variables, see the "Controlling Variables" section in the Skills Handbook.

▥▶ CHECK YOUR UNDERSTANDING

1. What could you call the container with no lemon juice? Why was it important to have this container?

2. Why was it important to have the same amount of yeast, sugar, and water in each container?

3. What was the dependent variable in this investigation?

Land Use and Habitat Loss

Humans use land in many ways. Human activities, such as farming, building cities, and logging, can change or eliminate the conditions that are needed for other living things to survive. Thus, human activities can result in loss of habitat for plants and animals. Land use issues are complicated, however. Looking after the needs of ecosystems and both the needs and wants of humans can be very difficult.

TRY THIS: *OBSERVE HABITAT LOSS*

Skills Focus: observing, inferring, classifying

Explain how each land use in the photos below has resulted in habitat loss. What species of plants and animals might have been affected? Do you think biodiversity has been reduced? Suggest some ways that habitat loss could be reduced in these land use situations.

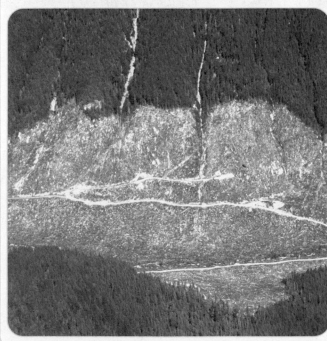

B4 • Vancouver Sun

Vineyards a threat to rare species

Larry Pynn Vancouver

A dry and little-known ecosystem—the antelope brush and needle-and-thread grass ecosystem—is being squeezed out by the ever-expanding grape-growing industry south of Penticton to the Canada–U.S. border. With the number of vineyards in the region predicted to double over the next several years, conservationists are urging immediate protection for what little antelope brush remains.

The antelope brush ecosystem is among the four most endangered habitats in Canada and is home to 88 species considered at risk by the B.C. government—a diverse and odd assortment of plants, insects, and animals, which, for the most part, are neither cute nor cuddly. Among the rarer species are the western red bat, tiger salamander, grasshopper sparrow, burrowing owl, white-tailed jackrabbit, badger, night snake, rough dropseed, flat-topped broomrape, Spalding's milk-vetch, Behr's hairstreak butterfly, and ground mantis.

Urbanization has been responsible for the loss of about 16 percent of antelope brush habitat. The other 84 percent has been lost to agriculture—vineyards, cultivated agriculture, orchards, and grazed pasture. Statistics are not specific to vineyards, but the consensus is that expansion of vineyards is the main threat.

The 400-member Osoyoos Indian Band (Nk'mip) has recently converted almost 500 ha (hectares) to vineyards as part of the band's mandate to become economically self-sufficient by 2005. Chris Scott, the band's chief operating officer, said he expects up to another 160 ha of antelope brush to be converted to vineyards over the next

Brian Sprout/Vancouver Sun

Conservationist Geoff Scudder in the endangered antelope brush system at the Osoyoos Desert Centre

two or three years, if the band membership approves. But he also said the band will be asked to endorse setting aside close to 400 ha of antelope brush habitat for conservation.

The B.C. Wine Institute argues that vineyards are trying their best to be environmentally responsible. Initiatives have included drip irrigation to reduce water use, netting and bird calls to keep out birds that would consume grapes, donations to wildlife conservation, and in some cases a movement to organic wines that are free of pesticides.

Jim Wyse, proprietor of Burrowing Owl Estate Winery in Oliver, said his company donates $20 000 a year to help recovery efforts for the burrowing owl. When an employee finds a western rattlesnake, it is relocated rather than killed. If a meadowlark nest is found in the vineyard, it is avoided until the young are fledged. And the vineyard has put out at least 60 nesting boxes for mountain bluebirds.

The antelope brush ecosystem is found in Canada only in the Okanagan and is considered "globally imperiled."

Many human activities cause habitat loss for other species (**Figure 1**). The case study shows how difficult it is to look after the needs of both humans and plants and animals in the same ecosystem. The needs and wants of British Columbians affect how we use the land in our province. Many of the ways that we use the land result in habitat loss for other species that live in British Columbia. Some of the food we eat and the products we use are produced in other provinces and countries. So our needs and wants affect land use in other provinces and countries. The choices we make here can result in habitat loss in other parts of Canada and in countries around the world.

LEARNING TIP ◁

It is easier to remember information if you personalize it. How do you feel about your choices concerning habitat loss? Compare your opinion with a partner.

Figure 1
Building this mall and its parking lot resulted in habitat loss in the local ecosystem. The production and transportation of the products sold in this mall resulted in habitat loss in ecosystems throughout Canada and in other parts of the world.

Canadians have a very high standard of living. We are able to meet many of our wants, as well as our needs. This means that we have a greater effect on land use and habitat loss than people in poorer countries with lower standards of living.

⫸ CHECK YOUR UNDERSTANDING

1. It is often difficult to look after the needs of both humans and ecosystems. Use examples from the article to explain why.

2. What type of land use causes the most habitat loss in the antelope brush ecosystem?

3. Using information from the article, explain how land use by humans can result in habitat loss for plants and animals.

4. The Nk'mip First Nation (near Osoyoos) is trying to protect rattlesnakes and their habitat in this ecosystem. Why do you think the Nk'mip consider rattlesnakes to be important?

5. Do human needs or human wants result in more habitat loss for other species? Explain your answer.

Perspectives on Land Use

Ecosystems are affected by the decisions that people make about land use. People have different values and needs, so they have different ideas about how to use the land. In this activity, you will work in a group to explore a land-use issue. Then you will role-play a roundtable discussion to reach a decision about the issue (**Figure 1**).

Figure 1

People share different viewpoints during a roundtable discussion. These different perspectives combine to help provide a solution to a problem.

▷ **LEARNING TIP**

To learn more about the steps to use in exploring an issue, see the "Exploring an Issue" section in the Skills Handbook.

The Issue

Identify a proposal for a change in land use that will affect local ecosystems. Here are some examples to start you thinking. You should identify a proposal from your own community, or a community near you.

- logging in a forested area
- opening up new land for agriculture

- developing a new landfill site
- opening up a natural area for new housing
- developing a recreational area, such as a ski hill
- setting aside a natural area as a park
- restoring streams in urban areas
- setting aside land for city gardens
- installing a basketball court in a neighbourhood park
- widening a highway

Background to the Issue

Identify Perspectives

Work in groups of four or five. Have each group member choose a role for the role-play. For example, you could be someone who is in favour of the land-use proposal, someone who is opposed, someone who has a few concerns, or someone who is not sure what to think. Use the following questions to help you identify the different perspectives:

- Why are people planning to change how the land is used?
- Are there people who think that the change is a good idea? How does the proposal meet their needs and values?
- Are there people who are opposed? How does the proposal affect their needs and values?
- Are there people who are not opposed but have a few concerns? What are their needs and values?
- Are there people who are undecided?
- Are there people who do not care? Why do they think the issue does not matter, or is it not important to them?

Gather Information

Research what the person you are role-playing is thinking and saying about the proposal. You may be able to get this information from news articles, the Internet, community groups, or Aboriginal Elders. Take careful notes so that you can accurately represent his or her perspective, whether or not you agree with it.

LEARNING TIP ◁

For help with researching, see the "Researching" section in the Skills Handbook.

www·science·nelson·com GO

Identify Possible Alternatives

When all the group members are ready, set up a roundtable. The table does not have to be round, but all the members should sit so that they face one another. Decide how you will take turns presenting your perspectives.

Role-play your chosen perspective. Be sure to suggest alternatives to the planned change in land use. These alternatives should not only meet the needs and values of the person you are role-playing, but also meet the needs and values of people with different perspectives.

Identify Criteria for Evaluating Solutions

Still in role at your roundtable, discuss to what extent your suggested alternatives

- meet the needs of community members
- respect the values of community members
- maintain biodiversity in the ecosystem

Make a Decision

Try to reach a decision at your roundtable on what the best course of action would be with respect to this land-use proposal.

Communicate Your Decision

As a group, write a brief report for your school newsletter or community newspaper. Explain your decision and the process you used to arrive at it.

⫸ CHECK YOUR UNDERSTANDING

1. Why is it important to consider different perspectives on an issue?
2. Why is it important to consider alternative solutions?
3. In what other situations might a roundtable discussion be useful?

Evidence for Climate Change

For years, scientists have predicted that climate change caused by global warming would first show up in the Far North. The people of Sachs Harbour on Banks Island in the Beaufort Sea know this is true from their own observations.

Rosemarie Kuptana (**Figure 1**) is from Sachs Harbour. Rosemarie's grandfather predicted that the icy ocean surrounding their home would get warm and the animals would suffer. Rosemarie now knows that both the scientists and her grandfather were correct. Climate change caused by global warming is easy to see in her community. Rosemarie is working to bring it to the attention of the world.

The Inuvialuit people of Sachs Harbour have been making observations about weather patterns in their area for much longer than the Canadian government has been collecting this information. They are therefore better able to identify any changes that are occurring. They have shared their indigenous knowledge with the scientists. They also helped make a video to warn the rest of the world about climate change.

On the video, the Inuvialuit people report early spring thaws and later autumn freeze-ups. They describe winter ice so thin it will not hold hunters—either humans or polar bears. They describe summer ice floes that are smaller now and farther out at sea, taking the seals that live on the ice floes beyond the reach of the people who need the seals for food. They show where the permafrost (permanently frozen ground) is melting, causing their house foundations to shift. They tell of finding sockeye salmon, which normally live 1500 km farther south, in the nets they set for Arctic char. They tell of seeing robins in the spring, a bird so rare this far north that there is no word for "robin" in their language. They tell of more frequent storms, which make fishing from boats more dangerous. They report experiencing thunder and lightning for the first time.

Figure 1
Rosemarie Kuptana organizes climate change observations made by the people of Sachs Harbour.

Rosemarie Kuptana took the video of her people's observations to a United Nations meeting on climate change in The Hague, Netherlands, in November 2000. Rosemarie is not only working to save her community. Melting Arctic ice will change weather patterns around the world, affecting all of us.

The Introduction of Non-Native Species

Non-native species are organisms that are living outside of their natural range. The introduction of non-native species is a lot like pollution. Non-native species can harm or destroy native species because they compete with them for food and water, or hunt them for food.

There are many examples of non-native species in British Columbia. **Table 1** lists several of them.

Table 1 Non-Native Species in British Columbia

Species		Description
bullfrog		large frog that was brought to British Columbia to be raised as restaurant food but was released when there was no market for frogs legs; eats native species of frogs, ducklings, garter snakes, songbirds, and even mice
domestic cat		pet that kills large numbers of songbirds
Atlantic salmon		non-native salmon that uses up spawning space and food supplies of Pacific salmon when it escapes from fish farms
purple loosestrife		garden plant that spreads and quickly takes over moist marshy areas; crowds out native plants that provide food and shelter for local animals
red-eared slider		pet that competes with endangered native species of turtles when it is released into the wild
smallmouth bass		popular fish with fishers, who have introduced it (illegally) into lakes and rivers in many parts of British Columbia; competes for food and habitat with native fish; eats juvenile trout and coho salmon fry

It is usually humans who introduce non-native species, either accidentally or on purpose. Non-native plants and animals can arrive on boats, mountain bikes, or hiking boots. They can escape from farms, gardens, and pet collections. Some people deliberately release their pets into the ecosystem because they think their pets will be happier or because they are tired of looking after their pets. Other people knowingly commit a crime by bringing wild animals from other countries here as pets. Fortunately, many people are involved in the attempt to control the introduction and spread of non-native species.

LEARNING TIP ◁

What is your reaction to what you have read about the introduction of non-native species? Check your understanding by comparing your reactions with a partner. Offer an opinion about what should be done about this problem.

Ⅲ▶ CHECK YOUR UNDERSTANDING ⊗

1. Use an example to explain how the introduction of a non-native species can decrease biodiversity in an ecosystem.

2. Suppose that a class in your school has been studying insects. In a science kit, the students obtained some insects called stick insects, sometimes called walking sticks (**Figure 1**). Stick insects normally live in a tropical climate. They reproduce quickly and eat leaves. One day, you notice that the window near the cage is open, and the stick insects are nowhere to be seen! Should you be alarmed? Explain why or why not.

Figure 1

3. Your friend has always loved animals and has three pets: a cat, a corn snake, and a budgie. Lately, she has decided that it is cruel to keep animals in cages. She says that she plans to release all her pets into the wild where they will be free and happy. Write a letter to her, explaining all the reasons why this might not be a good idea.

3.7 Working to Preserve, Conserve, and Restore Ecosystems

Many people are concerned about the negative effects that humans have on ecosystems. They want to do something to help. They want to be caretakers, or stewards, of ecosystems. Taking personal responsibility for looking after an ecosystem is called **stewardship.**

Preserving Ecosystems

More and more, people recognize the need to keep safe from destruction, or preserve, the remaining ecosystems that have not yet been damaged by humans. People are working to preserve some of the remaining natural areas in British Columbia, such as the antelope brush ecosystem in the Okanagan and the Khutzeymateen Valley. Ecosystems that have already been damaged by humans also need our stewardship, however.

Conserving Ecosystems

People are learning how to conserve, or make wise use of, existing ecosystems. For example, the Guichon family is trying to graze cattle on their ranchland in a way that preserves the natural grassland ecosystem (**Figure 1**). The Guichons move cattle between small areas of grassland, allowing the land time to recover after the cattle have grazed there. They also plan the movement of cattle between areas to provide birds, such as ducks and sharptail grouse, with nest sites.

Figure 1

The Guichon family won an award for their careful observations of the effects of cattle grazing on grasslands near Merritt.

Restoring Ecosystems

People often help to restore or repair the damage to ecosystems that was caused by previous human activities. An example of this is the Salmon River Restoration Project.

The Salmon River flows through forest, grasslands, and rocky gullies before it empties into Shuswap Lake. In the fall, different species of salmon travel from the ocean to the river to spawn.

Agriculture is the main human use of the area surrounding the river. Over the years, agricultural activities have changed the river, thus changing living conditions for the salmon. For example, water was taken from the river for irrigation. Logging and grazing cattle removed vegetation from the banks of the river, reducing shade and increasing water temperature. Water running off the fields carried fertilizers into the river. The runoff also carried soil into the river, silting up the gravel that salmon need for spawning beds.

Farmers, scientists, Aboriginal peoples, government officials, and volunteers are involved in the Salmon River Restoration Project (**Figure 2**). Students from nearby schools have helped plant trees and repair areas of the river. People with many different interests are working together to restore this ecosystem, using roundtable discussions.

LEARNING TIP ◁

Pause and think. Can you explain the difference between preserving, conserving, and restoring ecosystems in your own words? If not, re-read the section before you move on.

Figure 2
Cooperative landowners aid student volunteers in the restoration of the Salmon River.

Through preservation, conservation, and restoration projects, humans are having a positive impact on many ecosystems (**Figures 3 to 10**). These projects usually involve people from within the community and from outside. They all come together and work toward a common goal.

Figure 3
Steelhead are given a helping leap on the Bonaparte River.

Figure 4
Everett Crowley Park in Vancouver is on a former landfill site.

Figure 5
These bat houses were built by elementary students for bats that were left homeless. A fire destroyed their previous home in an old church.

Figure 6
The Kitlope Valley is a complete watershed, from its glacier peaks to the ocean floor. In 1994, a timber company gave up its rights to log this area.

Figure 7
Scientists and the Huu-ay-aht community (near Bamfield) are working together to bring back abalone stocks. Abalone are harvested as a food source and for their shells, which can be used in jewellery.

Figure 8
Thousands of wooden fish are shown in school and community "Stream of Dreams" fence murals. These murals remind local residents that their storm sewers are connected to a stream, river, or ocean.

Figure 9
To help burrowing owls, these volunteers are inserting large pipes into the ground in the grasslands ecosystem. The owls will be able to build their nests in the pipes.

Figure 10
Strathcona Community Gardens are located in East Vancouver. These city-sponsored gardens provide space for people to grow flowers and food.

TRY THIS: ***PRESERVE, CONSERVE, RESTORE***

Skills Focus: classifying

Look at **Figures 3 to 10**. Study the photos and read the captions. Decide whether each photo shows an attempt to preserve, conserve, or restore an ecosystem. Do some of the photos fit into more than one category? Discuss your decisions with a partner.

CHECK YOUR UNDERSTANDING

1. List three ways that people can work together to make a positive impact on ecosystems. Give an example of each, and state whether biodiversity is being maintained or increased.
2. Aboriginal groups are working with scientists and community groups on many of the preservation, conservation, and restoration projects in British Columbia. In a small group, brainstorm reasons why the involvement of Aboriginal groups is important.
3. Suggest at least one way in which people could have a positive impact on an ecosystem in your area.

Learning to Lessen Our Ecological Footprint

We use resources to meet our basic needs, to make our homes and clothing, and to provide energy for heating and transportation. We also use resources to provide a huge number of things that we do not need. We want these things, however, because they make our lives more comfortable, interesting, or enjoyable. How people choose to live has an effect on ecosystems.

On the other hand, there are many ways that people can restore the damage done to ecosystems. There are also many ways that people can behave to have a smaller negative impact on ecosystems. You can think of your impact as follows: When you put down your foot, you leave a footprint. In a similar way, when you use resources, you leave a footprint on the ecosystem in which you live (**Figure 1**). In 1996, Mathis Wackernagel and William Rees, two researchers at the University of British Columbia, developed a way to measure this ecological footprint, using summaries of resources used and waste produced. Their goal was to make us think about our ecological footprint as part of our stewardship of ecosystems.

Figure 1
How large is your ecological footprint?

> **LEARNING TIP**
>
> Make inferences—read between the lines. Ask yourself, "What does the ecological footprint idea really mean? Why is it important? How does it affect me?"

TRY THIS: DETERMINE YOUR FOOTPRINT

Skills Focus: collaborating, analyzing

Many Web sites can help you determine how big your ecological footprint is, or how much nature you use in your daily living. Search for the term "ecological footprint" on the Internet. Complete a survey to measure your ecological footprint. Compare your results with a classmate's results. What are the similarities and differences?

www·science·nelson·com

The David Suzuki Foundation in British Columbia is an organization that looks at how we can decrease our negative impact on nature (**Figure 2**). This organization has developed a list of ten ideas to help us use less nature in our daily living.

The Nature Challenge

Reduce home energy use by 10%.	Choose a fuel-efficient vehicle.
Choose an energy-efficient home.	Walk, bike, carpool, or take transit.
Replace dangerous pesticides.	Choose a home close to work or school.
Eat meat-free meals one day a week.	Support car-free alternatives.
Buy locally grown and produced food.	Learn more and share with others.

Figure 2
Have you ever asked what you can do to protect nature? David Suzuki has researched the ten most effective ways you can conserve nature. He challenges you to do three of them.

Canadians are encouraged to choose three of the ten ideas and commit to following them for one year. Which three ideas could you commit to following? If some of the ideas would be impossible for you, explain why. Is there one idea that your whole class could commit to following?

⫸ CHECK YOUR UNDERSTANDING

1. What does the term "ecological footprint" mean? Why is it important for people to reduce their ecological footprint?

2. List three ways that you could lessen your ecological footprint, without help from anyone else.

Chapter Review

Human survival depends on sustainable ecosystems.

Key Idea: Biodiversity keeps ecosystems strong and stable.

Vocabulary
sustainability p. 60
biodiversity p. 60

Key Idea: Indigenous knowledge can help us achieve sustainability.

"...respect, preserve and maintain knowledge, innovations and practices of indigenous and local communities embodying traditional lifestyles relevant for the conservation and sustainable use of biological diversity and promote their wider application with the approval and involvement of the holders of such knowledge, innovations and practices and encourage the equitable sharing of the benefits arising from the utilization of such knowledge, innovations and practices..."

Key Idea: Human activities can decrease biodiversity.

Pollution

Habitat loss

Introduction of non-native species

Key Idea: Humans can preserve, conserve, and restore ecosystems.

Vocabulary
stewardship p. 78

Preserve

Conserve

Restore

Key Idea: Humans can lessen their ecological footprint.

Review Key Ideas and Vocabulary

When answering the questions, remember to use vocabulary from the chapter.

1. Why is biodiversity important in an ecosystem?

2. What is a sustainable environment? How can indigenous knowledge help us achieve sustainability?

3. In point form, list some of the ways that human activities have reduced biodiversity in your local ecosystem.

4. Name and describe three ways that humans can work to have a positive impact on ecosystems.

5. What is an ecological footprint? Why does its size matter?

Use What You've Learned

6. How might building your school and schoolyard have resulted in habitat loss for plants and animals?

7. Imagine that you are a plant or animal in your local ecosystem. You are suffering because of pollution or loss of habitat. Write a descriptive paragraph from the point of view of this plant or animal.

8. Research the reasons why people felt that a grizzly bear sanctuary was needed in the Khutzeymateen Valley. Find out about other sanctuaries or wildlife preserves, and the reasons why they were established.

www·science·nelson·com

9. Why does the David Suzuki Foundation suggest eating meat-free meals one day a week? What is being conserved and how?

10. Choose one of the following situations. Create a cartoon, with captions, to show the effects this situation might have on your local ecosystem.
   a) A pet red-eared slider escapes.
   b) A person releases mink from a fur farm.
   c) A family moves to an apartment where pets are not allowed. They decide to let their pet cat live in the wild.

11. Abalone stocks are so low that it is illegal to harvest them off the coast of British Columbia. Use the Internet to research why abalone stocks are so low. What kind of restoration work is going on?

www·science·nelson·com

Think Critically

12. Does any other species besides human pollute? Explain your answer.

13. Research a conservation or preservation society in British Columbia. Determine what the goals of this society are. Decide whether you would support this society, and give reasons for your decision.

14. Do you think that the ecological footprint of the average Canadian is larger or smaller than the ecological footprint of someone who lives in a less developed country, such as Haiti? Why?

Reflect on Your Learning

15. Describe how something you learned in this chapter will affect your future behaviour.

16. What topic or issue in this chapter would you like to investigate further? What questions do you still have? How will you find answers to these questions?

Making Connections

Taking Action in Your Local Ecosystem

Before you take action in your local ecosystem, gather information from many possible sources. Don't forget to include indigenous knowledge.

Looking Back

In this unit, you have learned how ecosystems support life. You have learned about the complexity of interactions and cycles. You now know how easily the web of life can be damaged by human activities. You also know how important it is to be a steward, or caretaker, of your ecosystem.

In this activity, you will demonstrate what you have learned as you exercise stewardship in your local ecosystem. You will work with other students to solve a problem in your local ecosystem.

Demonstrate Your Learning

Part 1: Identify a problem.

Is there a problem in your local ecosystem that you could try to solve? Make sure that the problem you choose is a manageable size for you. Here are some suggestions:

- loss of native plants
- toxic herbicides being used to control weeds
- improper disposal of hazardous wastes
- idling engines
- backyard burning

- electronic waste
- erosion from stormwater runoff
- invasive species
- damaged shorelines
- low participation in recycling programs
- overpackaging of consumer products
- damage to ecosystems from recreational activities, such as snowmobiling

Define the problem as precisely as you can.

Part 2: Define the task.
Decide at which level you are going to work—in your classroom or school, or in your community. Then decide on an approach to solve your problem. You can choose one of two basic approaches: direct action or education. If you choose to take direct action, then you will be actively involved in planning a project or helping complete one. If you choose to educate people, then you must research your issue and find a persuasive way to present it.

Part 3: Define criteria for success.
How would you define success in ecosystems? How would you define success in your project?

Part 4: Gather information.
Problems in ecosystems are complex, so it is important to gather background information. Make a list of questions that you will need to get answered. Also make a list of possible sources of information. Remember to include indigenous knowledge.

Part 5: Plan.
Make a list of specific steps. Decide when you will work on each step. Your steps may include getting permission to do certain things. Make a list of all the materials you will need, and where you will get them.

Part 6: Test.
Test your solution against the criteria you established. If you plan to make a direct change in an ecosystem, test the change in a small area first.

Part 7: Evaluate.
Did your solution go as planned? If not, why do you think it turned out differently than you planned? Did you meet your criteria for success? If not, what could you have done differently?

Part 8: Communicate.
Communicate your results, as well as any recommendations you have for further actions, to an appropriate audience. An appropriate audience may be your classmates, the general public, an organization, or a government agency.

ASSESSMENT

Check to make sure that your work provides evidence that you are able to

- evaluate human impact in a local ecosystem
- analyze the roles of organisms in a local ecosystem
- assess the survival needs of organisms
- assess the interactions between organisms and the non-living parts of their environment
- assess the requirements for maintaining or restoring biodiversity

CHAPTER
4 Matter can be described using properties.

CHAPTER
5 Matter is made up of moving particles.

CHAPTER
6 Matter can be classified.

Preview

The beautiful frost crystals in the photograph might inspire a poet. They might also inspire a scientific mind. You probably know that frost is made of frozen water, but how is it formed? Under what circumstances can frost form large crystals like the ones shown here? How would you describe these frost crystals poetically? How would you describe them scientifically?

In this unit, you will learn about matter (such as frost) and ways to describe matter scientifically. You will observe, measure, and compare many different types of matter. You will learn how to safely conduct investigations using matter and how to communicate your results. Investigations are extremely important to the study of chemistry. You must be able to follow procedures carefully, handle chemicals safely, make detailed observations, and come to conclusions.

To help you understand your observations, you will learn about a model that explains the behaviour of matter. You will find out how scientists classify matter and how physical and chemical changes affect your daily life. By the end of this unit, you will have a greater understanding and appreciation of the influence of chemistry on the world around you.

TRY THIS: MAKE A KWL CHART

Skills Focus: communicating

Before you start this unit, make a three-column K-W-L chart. Record what you already know about chemistry in the first column. Record what you wonder about chemistry in the second column. After you finish the unit, write some things you learned in the third column.

Know	Wonder	Learned

◀ Frost crystals on a windowpane.

CHAPTER 4

Matter can be described using properties.

KEY IDEAS

▸ Matter can be described using observable properties.

▸ Matter can be described using measurable properties.

▸ Matter is anything that has mass and volume.

▷ **LEARNING TIP**

As you read the first two paragraphs, try to answer the questions using what you already know.

Eulachons (or oolichans) are sometimes called candlefish because a dried eulachon can be burned like a candle. They are also called oilfish because their bodies are 20% oil by weight. For hundreds of years, many Aboriginal peoples have collected and used the oil—often called grease—from eulachons. How would you describe the eulachon grease in the photograph? Is it a solid or a liquid? What colour is it?

Aboriginal peoples use eulachon grease to season food, preserve fruit, and lubricate tools. They also use it as a medicine. They had practical knowledge of the chemical characteristics or properties of eulachon grease that allow it to be collected and used in these ways. Scientists explain the scientific principles behind people's working knowledge of things. In this chapter, you will learn how all matter—from eulachon grease to the air you breathe—can be described scientifically.

Properties of Matter

When you choose your clothes, your lunch, and even your toothpaste, you are making choices based on the properties of matter. A **property** is a characteristic that may help to identify a substance. You can observe properties using your five senses, or you can determine properties using simple tests and measurements.

Properties You Can Observe with Your Senses

You can use your five senses—sight, touch, hearing, smell, and taste—to observe matter (**Figure 1**).

LEARNING TIP

Before reading this section, "walk" through it and make a note of the headings and subheadings. Use these to take point form notes as you read.

Figure 1
Which senses would you use to describe the properties of an ice-cream sundae?

Some of the properties you can observe with your senses are summarized in **Table 1**.

Table 1 Properties Observed with the Senses

Property	Describing the property
colour	Is it black, white, colourless, red, blue, greenish-yellow …?
taste	Is it sweet, sour, salty, bitter …?
texture	Is it fine, coarse, smooth, gritty …?
odour	Is it odourless, spicy, sharp, burnt …?
lustre	Is it shiny, dull …?
clarity	Is it clear, cloudy, opaque, translucent …?

States of Matter

You can also use your five senses to observe whether a substance is a solid, a liquid, or a gas. These are called the **states** of matter. A substance may be found in all three states. For example, water can be found as a solid (ice), a liquid (water), or a gas (water vapour in the air), depending on the temperature. You can easily observe the state of a substance at room temperature.

TRY THIS: OBSERVE PROPERTIES

Skills Focus: observing, communicating

Play "I spy" with a partner using the observable properties of matter. Use the format, "I spy something that is (pick a state) and is (pick one or more properties from **Table 1**) …" For example, "I spy something that is a solid, and is blue and shiny. What is it?"

Check with your teacher before you taste anything other than your own lunch.

Properties You Can Measure

Some properties can be determined using simple tests and measurements. For example, you could put a substance in water to see if it dissolves. You could also put a variety of substances in water to see which ones float and which ones sink. Later in this unit, you will measure properties of matter using both of these tests.

Melting and Boiling Points

One of the properties of matter that can be measured is the temperature at which a substance changes state. Most substances have two temperatures at which they change state.

The **melting point** of a substance is the temperature at which the solid form of the substance changes to a liquid (**Figure 2**). For example, water changes from solid ice to liquid water at 0°C. Thus, the melting point of solid water (ice) is 0°C.

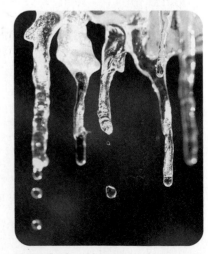

Figure 2
The melting point of ice is 0°C.

The **freezing point** of a substance is the temperature at which the liquid form changes to a solid. If water is cooled, it will freeze at 0°C. The freezing point of a substance is the same as its melting point.

The **boiling point** of a substance is the temperature at which the liquid form of the substance changes to a gas. For example, at the boiling point of water, 100°C, liquid water changes to water vapour, a gas (**Figure 3**).

Figure 3
The boiling point of water is 100°C.

Melting point and boiling point are properties that can be used to help identify a substance.

▶ CHECK YOUR UNDERSTANDING

1. Make a chart, like the one below, that lists properties you can observe using your senses and properties you can observe using simple measurements.

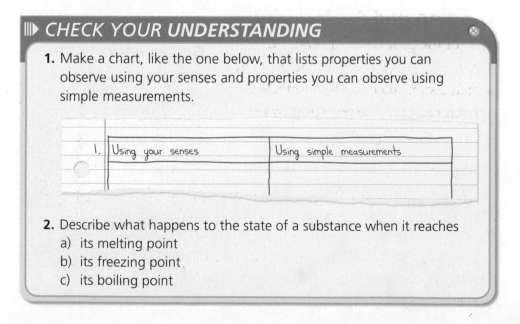

2. Describe what happens to the state of a substance when it reaches
   a) its melting point
   b) its freezing point
   c) its boiling point

4.1 Properties of Matter　**93**

▷ **LEARNING TIP**

To review line graphs and writing a hypothesis see the Skills Handbook sections "Graphing Data" and "Hypothesizing."

Ice to Water to Steam

Suppose that you leave an ice cube at room temperature (20°C). Heat from the surrounding air will melt the ice and turn it into water. Then, if you heat the water enough, it will boil and change into water vapour. In this investigation, you will explore what happens to the temperature of water as it changes state (**Figure 1**).

Figure 1

In winter, you can easily find water in all three states. Look at this photo of Pine Creek Falls in northern British Columbia. Solid water is found as ice and snow, liquid water runs under the ice, and some gaseous water (water vapour or steam) is always present in the air.

Question

What will happen to the temperature of water as it changes state? Make a prediction by drawing a line graph of temperature versus time. Put temperature on the y-axis and time on the x-axis. Use your graph to predict what you think will happen to the temperature of water as it is heated from ice to liquid to water vapour. Make sure that your graph includes any important temperature values.

Hypothesis

Write a hypothesis based on your prediction. Use the form "If … then …."

crushed ice

Pyrex beaker

stirring rod

thermometer

watch

hot plate

stand and clamp

Materials

- 250 mL crushed ice
- 250-mL Pyrex beaker
- stirring rod
- thermometer
- watch or clock that displays minutes and seconds
- hot plate
- stand and clamp apparatus

Part 1: Ice to Water

1 In your notebook, draw a table like the one below.

Data Table for Investigation 4.2

Time (min)	Temperature (°C)	Other observations
0		
1		
2		
3		
4		
5		
6		
7		
8		
9		
10		
11		
12		
13		
14		
15		
16		
17		
18		
19		

Thermometers break easily. Remove the thermometer from the beaker before stirring the ice. Do not let the thermometer touch the bottom of the beaker. Do not leave the thermometer in the beaker, as it may be too heavy and tip the beaker.

2 Place the crushed ice in the beaker. Stir the ice with the stirring rod. Place the thermometer in the beaker, and measure the temperature. Record this temperature in your table as the temperature at 0 min. Remove the thermometer. Record your observations of any changes you see in the right-hand column of your table.

3 After 1 min, stir the ice again. Then measure and record the temperature. Record your observations of any changes you see in the right-hand column of your table.

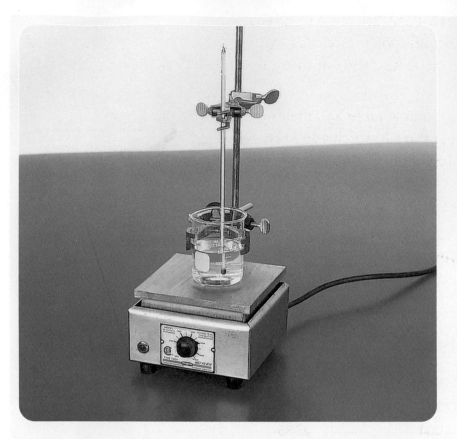

4 Repeat step 3 every minute, until 5 min after all the ice has melted. Record your observations of any changes you see in the right-hand column of your table.

Part 2: Water to Steam (Teacher Demonstration)

5 To find out what happens to the temperature of water as it boils, your teacher will do a demonstration, as shown on the left. Record the temperature every minute as the water is heated and for at least 5 min after it begins to boil. Record your observations of any changes.

Analyze

1. Use your data to create a line graph of temperature versus time. Describe the shape of your graph.

2. Describe how temperature changes as ice melts. Describe how temperature changes as water boils.

3. Predict what your graph would look like if you could continue to heat the water vapour.

Write a Conclusion

4. How did your predicted graph compare with your actual graph? Was your hypothesis correct? Why or why not? Did your observations support, partly support, or not support your hypothesis? Write a conclusion for your investigation.

Apply and Extend

5. Based on your observations, do you agree with the following statement? Explain your answer.

"When heat is added to a solid, it can cause a change of state or an increase in temperature."

6. Suppose that you are camping in the fall. You leave some water in a pail overnight. The next morning, you notice a layer of ice on the top of the water. What is the temperature of the water just below the ice?

7. In section 4.1, you learned about boiling point and melting point. Melting point is the temperature at which a solid changes to a liquid. Boiling point is the temperature at which a liquid changes to a gas. **Table 2** lists the boiling points and melting points of some common substances.

   What was the melting point of your ice in this investigation? What was the boiling point of your water? If your values are different from those in **Table 2**, what are some possible reasons for the difference?

Table 2 Melting Points and Boiling Points of Some Common Substances

Substance	Melting point (°C)	Boiling point (°C)
ethanol	−114	78
copper	1084	2336
oxygen	−218	−183
sodium chloride (table salt)	801	1465
sulfur	113	445
water	0	100

⫸ CHECK YOUR UNDERSTANDING

1. Why did you need to stir the ice-water mixture?
2. Where did you put the bulb of the thermometer to get the most accurate reading? Why?
3. Why was it important to measure the temperature at regular intervals?

Plasma

A gas that has electricity running through it is called plasma. Plasma is sometimes considered to be a fourth state of matter. It is found mainly in the stars and nebulas within our universe.

Figure 1
The northern lights (aurora borealis)

Figure 2
Neon lights contain plasma.

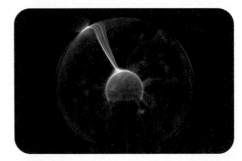

Figure 3
You can see plasma balls at science centres or science stores.

Plasma has fascinated people for thousands of years. The northern lights (aurora borealis) are an example of plasma in nature (**Figure 1**). In ancient times, the Inuit people believed that the northern lights were the torches of spirits guiding souls to a land of happiness and plenty.

Today you can find plasma in many manufactured items, such as fluorescent lights, neon signs (**Figure 2**), and plasma balls (**Figure 3**). The wonder that you experience when looking at a plasma ball is like the wonder the ancient Inuit people experienced when looking at the northern lights.

Plasma can even be used to cut and shape metal. Plasma cutters (**Figure 4**) were developed almost 50 years ago, during World War II, to help speed up the process of cutting and welding metal together to build airplanes for the war. Plasma cutters are now used to shape car frames, to cut large beams of metal at construction sites, and are even used by artists to cut and shape metal for sculptures.

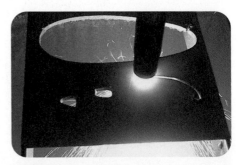

Figure 4
A plasma cutter cutting metal.

Television screens are one of the newest technologies that involve plasma. Plasma displays are not a new invention—research on plasma displays dates back a decade or more. It is only recently, however, that the technology has been developed to manufacture plasma displays at a lower cost.

So how does a plasma screen television work? A plasma screen is quite different from a regular television screen. A plasma screen works by suspending an inert (inactive) gas, such as neon or xenon [ZEE-non], between two panes of transistor-covered glass that are meshed together. An electric charge is applied to the gas, turning it into plasma. This creates ultraviolet light. The ultraviolet light illuminates phosphors that are built into the glass, creating light that you can see (**Figure 5**).

In less scientific terms, think about one million very small light bulbs arranged between two glass plates. The light bulbs are lit by plasma and produce the spectrum of colour needed to view an image. The light bulbs are turned on or off by the television's processor.

Why do so many people wish to have a plasma screen television? Plasma screen televisions are so thin and light that they can be hung on a wall (**Figure 6**). This means you can get a clear view of a plasma screen from almost any angle in a room. As well, there is very little reflection off a plasma screen. These characteristics make plasma screen televisions very desirable.

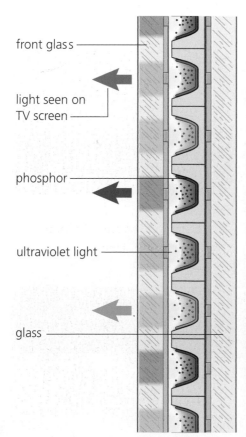

front glass

light seen on TV screen

phosphor

ultraviolet light

glass

Figure 5
How a plasma screen works

Figure 6
Plasma screen televisions are very light and very thin.

Mass and Volume

In this chapter, you are investigating some of the properties of matter. Everything in the world, including you, is made of matter. What exactly is matter? **Matter** is anything that has mass and occupies space. In this section, you will learn about mass and volume.

Mass

The **mass** of an object is a measure of the amount of matter in the object. An object's mass stays constant everywhere in the universe.

Mass is used to measure many things, from food to people (**Figure 1**). For example, when you buy a bag of potato chips, you are buying a certain mass of potato chips. Small masses, such as the mass of a bag of potato chips, are often measured in grams. Larger masses, such as the mass of people or vehicles, are often measured in kilograms (*kilo* means "1000"). Very small masses, such as the amounts of some medicines, are measured in milligrams (*milli* means "one-thousandth," or "$\frac{1}{1000}$").

$$1 \text{ mg} = \frac{1}{1000} \text{ g}$$

$$1 \text{ kg} = 1000 \text{ g}$$

Figure 1

The mass of objects with different amounts of matter can be measured in different units.

Prefix	kilo		centi	milli
Multiple	1000		$\frac{1}{100}$	$\frac{1}{1000}$
Length	kilometre (km)	**metre (m)**	centimetre (cm)	millimetre (mm)
Mass	kilogram (kg)	**gram (g)**	centigram (cg)	milligram (mg)
Volume		**litre (L)**		millilitre (mL)

LEARNING TIP ◁

The International System of Units, or metric system, is commonly referred to as SI. SI comes from the French name, *Le Système internationale d'unités.*

◀

Measuring Mass

When you measure the mass of an object on a balance or a scale, you are measuring the mass directly. Therefore, this is an example of direct measurement.

Sometimes, you need to use a more complicated method to measure mass. For example, to find the mass of a quantity of water, you first need to find the mass of an empty, dry container. Then you pour the water into the container and find the mass of the container and the water. Finally, you subtract the mass of the empty container from the mass of the container with the water in it. The formula is

Mass of water = (mass of container + water) − mass of container

This is an example of indirect measurement.

Volume

As well as having mass, matter occupies space. **Volume** is a measure of the amount of space that is occupied by matter.

Measuring the Volume of a Liquid

You can measure a small volume of a liquid directly in a graduated cylinder. A graduated cylinder is a tall, narrow container with a scale of numbers on the side (**Figure 2**).

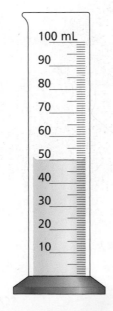

Figure 2
A graduated cylinder is marked out in steps (graduations) to enable measurement.

To measure the volume of a liquid in a graduated cylinder, you read the scale of numbers. When you look at a liquid in a graduated cylinder from the side, you will notice that the top surface has a slight curve where the liquid touches the cylinder. This curved surface is called the meniscus. For an accurate measurement, you should have your eye level with the meniscus, as shown in **Figure 3**. Then you read the volume at the bottom of the meniscus.

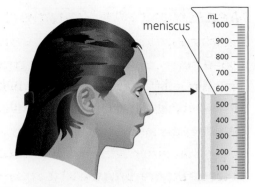

Figure 3
Read the volume of a liquid from the bottom of the meniscus.

The volume of a liquid is generally measured in litres (L) or millilitres (mL). (A millilitre is $\frac{1}{1000}$ of a litre.) You will be familiar with measurements of volume from containers of milk or soft drinks.

Calculating the Volume of a Rectangular Solid

You can measure a rectangular solid with a ruler and then calculate its volume using the following formula:

Volume = length $\times$ width $\times$ height

If you measure all the sides in centimetres, then the volume will be in cubic centimetres (cm^3). If you measure all the sides in metres, then the volume will be in cubic metres (m^3).

The volume of a rectangular solid with a length of 3 cm, a width of 4 cm, and a height of 2 cm (**Figure 4**) is calculated as follows:

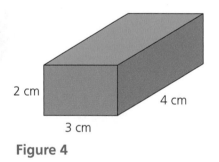

2 cm

3 cm

4 cm

Figure 4

Volume = length $\times$ width $\times$ height
= 3 cm $\times$ 4 cm $\times$ 2 cm
= 24 cm^3

The volume of a solid is usually given in cubic centimetres (cm^3). The volume of a liquid is usually given in millilitres (mL). Recipes, however, usually use millilitres for both solid and liquid volumes. This works because 1 cm^3 is the same as 1 mL, and 1000 cm^3 is the same as 1 L. Thus, in the calculation above, the volume could also be stated as 24 mL.

Measuring the Volume of an Irregular Solid

Sometimes, you cannot measure the length, width, and height of a solid because the sides are not regular. The volume of a small, irregular solid, such as a jagged rock, must be measured by displacement. To do this, choose a container (such as a graduated cylinder) that the irregular solid will fit inside. Pour water into the empty container until it is about half full. Record the volume of water in the container, and then carefully add the solid. Make sure that the solid is completely submerged in the water. Record the volume of the water plus the solid (**Figure 5**). Calculate the volume of the solid using the following formula:

Volume of solid = (volume of water + solid) − volume of water

LEARNING TIP ◁

For review in measuring mass and volume, see "Measurement and Measuring Tools" in the Skills Handbook.

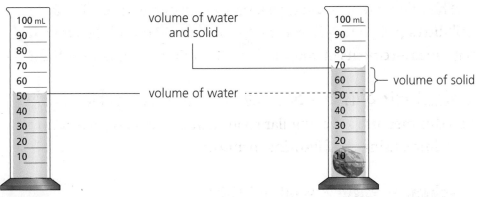

Figure 5
Measuring volume by the displacement of water

⫸ CHECK YOUR UNDERSTANDING

1. Define mass and volume. State the units that are used to measure each.

2. Name two pieces of equipment that can be used to measure mass and volume. How do these pieces of equipment improve our ability to communicate with each other?

3. What is the correct way to read the volume of a liquid in a graduated cylinder?

4. Determine the volume of the following box.

6 cm

5 cm

10 cm

4.4 *Conduct an Investigation*

● SKILLS MENU

○ Questioning	● Observing
● Predicting	● Measuring
○ Hypothesizing	○ Classifying
○ Designing Experiments	○ Inferring
○ Controlling Variables	● Interpreting Data
○ Creating Models	● Communicating

Measuring Mass and Volume

In this investigation, you will use what you learned in section 4.3 to determine the mass and volume of some common classroom objects (**Figure 1**). First you will estimate the mass and volume of these objects. Then you will check your estimates using direct measurement or the displacement of water.

Figure 1
How would you determine the mass and volume of these objects?

Question

What is the mass and volume of common classroom objects?

Materials

- safety goggles
- variety of regular solids (for example, textbook, dice, and block of wood)
- variety of small irregular solids (for example, small rock, small spoon, and metal nut)
- balance or scale
- ruler
- 100-mL graduated cylinder or large measuring cup
- water

Procedure

1 Estimate the mass of each object in grams. Record your estimates in your notebook in a table like the one below.

Data Table for Investigation 4.4

Object	Estimated mass (g)	Actual mass (g)	Estimated volume (cm³ or mL)	Actual volume (cm³ or mL)
textbook				
eraser				

2 Use the balance or scale to determine the actual mass of each object in grams. Record your results in your table, under "Actual mass."

3 Estimate the volume of each object in either cm³ or mL. Record your estimates in your table.

4 Determine the actual volume of each rectangular solid in cm³. Record your measurements, calculations, and results on your table, under "Actual volume."

5 Determine the actual volume of each irregular solid in mL using displacement. Remember to tilt the graduated cylinder or measuring cup and gently slide the solid into the water.

Record your measurements, calculations, and results in your table, under "Actual volume."

Analyze and Evaluate

1. Which masses or volumes were you able to estimate most accurately? Why?

2. Which masses or volumes did you estimate least accurately? Why?

3. You used the displacement of water to measure the volumes of irregular solids.
   a) Explain why "displacement of water" is an appropriate name for this method.
   b) Why is this method an example of indirect measurement?

Apply and Extend

4. Describe two everyday situations in which the measurement of mass or volume is important.

5. Imagine that you are provided with a scale, a sample of modelling clay, a piece of string, a graduated cylinder, and some water (**Figure 2**). How could you use these materials to prove that you can change the shape of the clay without changing the volume of the clay?

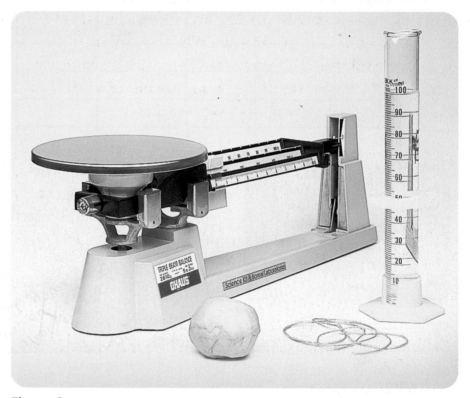

Figure 2

⫸ CHECK YOUR UNDERSTANDING

1. Why did you slide each object into the graduated cylinder rather than dropping it in? Would your results have changed if you had not slid all the objects into the cylinder in the same way? Would you have still obtained fair measurements? Explain your answer.

2. When would the displacement of water not be a good method for finding the volume of an irregular object?

Calculating Density

Look at **Figures 1** and **2**. In both photos, the oil is floating on the water. This property of oil makes it possible to clean up an oil spill and to skim the oil from a boiling pot of eulachons [YOO-luh-kons]. Why does oil float? Oil must be lighter than water, but what does this mean? A litre of oil is certainly not lighter than a glass of water.

To compare fluids using the words "light" and "heavy," you must examine the same volume of each fluid. Thus, a litre of oil is lighter (has less mass) than a litre of water. When you compare the masses of the same volume of different substances, you are comparing their densities. **Density** is the mass per unit volume of a substance. Oil floats on water because it is less dense than water.

Figure 1
An oil spill being contained.

LEARNING TIP ◁

Make connections to your prior knowledge. Ask yourself, "What do I already know about floating and sinking? How does this information fit with what I already know?"

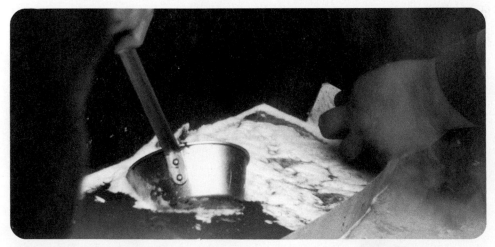

Figure 2
Eulachon oil being skimmed from a pot.

Skills Focus: observing, predicting, measuring

1. Find six identical opaque containers, such as plastic film containers. Fill the containers with different materials, such as water, sand, tiny pebbles, syrup, shampoo, and wood chips (**Figure 3**).

2. Close the containers and mix them up so you do not know which one is which. Number the containers.

3. Rank the containers in order from highest density to lowest density. You may use any method you choose to determine your ranking, but you cannot open the containers. Record your ranking.

4. Which densities were you able to estimate quite accurately? Which were harder to estimate? Why?

Figure 3

Using Density

Density is a property of matter that can be calculated. It is the mass of a substance per unit volume of this substance. It is expressed as grams per cubic centimetre (g/cm^3) or grams per millilitre (g/mL).

Density is calculated by dividing the mass of an amount of substance by its volume. The formula is

$$Density = \frac{mass}{volume}$$

Each substance has its own unique density. Water has a density of 1.0 g/mL. Liquids and solids that float on water have a density of less than 1.0 g/mL. Liquids or solids that sink in water have densities of more than 1.0 g/mL.

Table 1 lists the densities of some common substances. Notice that western red cedar has a lower density than water. Therefore, western red cedar floats in water, as do most types of wood (**Figure 4**). Crude oil also has a lower density than water, which is why oil spills stay afloat in the ocean. Copper has a higher density, however, so it sinks in water. The densities of two substances can be used to predict which will float and which will sink.

LEARNING TIP ◁

Make connections to your prior knowledge. Ask yourself, "How does this information on density fit with what I already know?"

Table 1 Densities of Some Common Substances

Substance	Density (g/mL)
wood (western red cedar)	0.37 (approximate)
crude oil	0.86–0.88 (approximate)
pure water	1.00
copper	8.92

Figure 4
What property of wood allows forest companies to transport logs in this way?

▌▶ CHECK YOUR UNDERSTANDING

1. What is density? How is it calculated?

2. Use the "Actual mass" and "Actual volume" columns of your data table for Investigation 4.4 to calculate the density of each object.

3. Calculate the density of each kind of wood.
   a) a child's block made of birch wood with a volume of 510 cm³ and a mass of 306 g
   b) a pine log with a mass of 96 000 g and a volume of 240 000 cm³
   c) a sculpture made of ebony with a volume of 81 cm³ and a mass of 96 g

Will It Float or Sink?

Will a rock float in water? Will a cork float in alcohol? Will alcohol float on glycerine? In this investigation, you will predict whether various solids will float or sink in three liquids. Then you will test your predictions. You will also predict and test what will happen when you combine the three liquids.

Question

Which materials will float or sink in rubbing alcohol, water, and glycerine?

Materials

- safety goggles
- apron
- small pieces of various solids (such as cork, wood, and rock)
- ruler
- graduated cylinder
- balance
- 3 250-mL beakers or small glass jars
- rubbing alcohol (isopropyl alcohol, density 0.8 g/mL)
- water (density 1.0 g/mL)
- glycerine (density 1.3 g/mL)
- 3 colours of food colouring

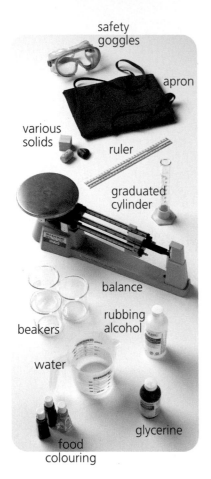

safety goggles
apron
various solids
ruler
graduated cylinder
balance
beakers
rubbing alcohol
water
glycerine
food colouring

▶ SKILLS MENU

○ Questioning	● Observing
● Predicting	● Measuring
○ Hypothesizing	● Classifying
○ Designing Experiments	○ Inferring
○ Controlling Variables	● Interpreting Data
○ Creating Models	● Communicating

▶ Procedure

1 Calculate the density of each solid. You may have to use indirect measurement to determine the volumes of some of the solids.

Record the densities in your notebook in a table like the one below.

Data Table for Investigation 4.6

Materials tested	Mass	Volume	Density (g/cm³)	Will it float or sink ...?		
				in rubbing alcohol	in water	in glycerine
ice						
wax						

2 Use your densities to predict which solids will float in each liquid. Write your predictions as "yes" or "no" in your table.

3 Put on your apron and safety goggles.

 Rubbing alcohol can harm your eyes. Wear safety goggles at all times.

solid in the three liquids. Use a check mark (✔) or an (✘) to indicate whether or not each prediction is correct.

5 Pour half of the water out of your beaker. Use a drop of food colouring to make each liquid a different colour. In your notebook, predict what will happen if you combine the three liquids. **Gently** pour some of the alcohol and then some of the glycerine into the water. Record your observations.

4 Fill each beaker three-quarters full with one of the liquids. Test your predictions by placing each

Analyze and Evaluate

1. Summarize your results in a few sentences.

2. Explain how you can use density to predict whether or not one substance will float on another substance.

Apply and Extend

3. Which substance that you tested is the most dense? Which is the least dense? Give one use for each substance that relies on its density.

4. In the last step of the procedure, you combined different liquids. Describe the final appearance of the combined liquids. What can you conclude about the densities of the three liquids?

5. **Table 2** gives the densities of several metals. Mercury is the only metal that is a liquid at room temperature. Mercury is very toxic. You should never touch it or inhale its vapours. Use **Table 2** to determine which metals would float and which would sink in liquid mercury (**Figure 1**).

Figure 1
Mercury is a silvery-white, liquid metal.

Table 2 Densities of Some Common Metals

Metal	Density (g/mL or g/cm³)
aluminum	2.7
chromium	7.2
copper	8.95
gold	19.3
iron	7.86
lead	11.34
mercury	13.6
silver	10.5
tin	7.31
zinc	7.13

⫸ CHECK YOUR UNDERSTANDING

1. How could mistakes in your measurements or calculations have affected the accuracy of your predictions?

2. Could the food colouring you added to the liquids in step 5 have affected the densities of the liquids? Explain.

Matter can be described using properties.

Key Idea: **Matter can be described using observable properties.**

WHAT WILL MY SUNDAE TASTE LIKE? WHAT DOES IT SMELL LIKE? WHAT COLOUR IS IT? WILL IT BE SMOOTH?

You can use your senses to describe the colour, texture, and state of this sundae.

Vocabulary

property p. 91

states p. 92

Key Idea: **Matter can be described using measurable properties.**

Melting point

Boiling point

$$\text{Density} = \frac{\text{mass}}{\text{volume}}$$

Density

Vocabulary

melting point p. 92

freezing point p. 93

boiling point p. 93

density p. 107

Key Idea: **Matter is anything that has mass and volume.**

• Mass is a measure of the amount of matter in an object.

• Volume is a measure of the amount of space that is occupied by matter.

Vocabulary

matter p. 100

mass p. 100

volume p. 101

Review Key Ideas and Vocabulary

When answering the questions, remember to use vocabulary from the chapter.

1. List properties that you can observe using only your senses. Choose an object in your classroom, and describe it using these properties.

2. Name two properties that require measurements.

3. Describe both the equipment you would need and the steps you would take to measure
   a) the volume of a ring
   b) the mass of a sample of liquid
   c) the volume of a cement block
   d) the mass of a stone
   e) the volume of a sample of liquid

4. Define density. Why is density considered to be a property of matter, but length is not?

Use What You've Learned

5. Vinegar and water are both clear liquids at room temperature (**Figure 1**). What properties could you use to tell them apart?

Figure 1

6. If a substance is a solid at room temperature (20°C), what can you say about its melting point?

7. Look at the melting and boiling points of mercury and ethanol (**Table 1**). Which substance would be better to use in an outdoor thermometer in the Arctic? Why?

Table 1 Melting Points and Boiling Points of Two Substances

Substance	Melting point (°C)	Boiling point (°C)
mercury	−38.9	356.6
ethanol	−114.3	78.5

8. For each substance, state which is the larger value.
   a) 340 mL or 1 L of apple juice
   b) 100 g or 0.5 kg of laundry soap
   c) 50 L or 500 mL of water in a bathtub
   d) 2 kg or 500 g of potatoes

9. The mass of a dry, empty beaker is 250 g. The mass of the beaker and a liquid is 475 g. What is the mass of the liquid?

10. A classroom measures 11.0 m by 9.0 m by 3.0 m. What is the approximate volume of air in the classroom?

11. A graduated cylinder contains 40 mL of water. A stone is carefully slipped into the cylinder. The level of the water reaches 57 mL. What is the volume of the stone?

12. a) Suppose that you tried to use the displacement of water to find the volume of a sugar cube. What problem could you have? What could you do to solve this problem?
    b) List two other objects whose volumes you could not measure using the displacement of water. Explain why.

13. Describe a method you could use to determine the volume of your body.

14. An ice cube is placed on one balance pan of an equal-arm balance. Masses totalling 3.5 g are placed on the opposite pan to level the balance (**Figure 2**). If the ice cube is allowed to melt, do you expect the balance to stay level? Explain.

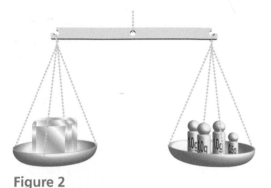

Figure 2

15. Investigate methods that are used by Aboriginal peoples to obtain the oil from eulachons (**Figure 3**). Identify the properties of oil that are used in these methods.

www·science·nelson·com **GO**

Figure 3

16. A hydrometer can be used to measure density of liquids (**Figure 4**). Conduct research to find out what a hydrometer is and how it measures density.

www·science·nelson·com **GO**

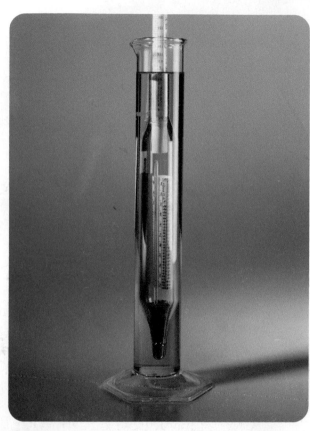

Figure 4

Think Critically

17. During a class discussion, one student states that solids are always denser than liquids. Several other students disagree with this statement. Which position would you take? Give examples to support your position.

Reflect on Your Learning

18. You have learned a lot about matter in this chapter. Think back to the beginning of this chapter. How have your ideas about matter changed?

Matter is made up of moving particles.

Imagine that you are sitting around a large campfire with some friends. You put a pot of water over the campfire. When the water begins to bubble and steam, you add some powdered hot chocolate mix and stir until the powder dissolves. Meanwhile, your friends are toasting wieners and marshmallows. Suddenly, one marshmallow catches fire and burns brightly for an instant. Your friend blows out the flame and looks at the black crispy chunk that is left on the stick.

Several changes took place around this campfire. Water changed state from a liquid to a gas. You made a drink by mixing the water with a powder. A marshmallow underwent some type of change and turned black, but remained a solid.

In this chapter, you will learn about a model that explains the behaviour of matter. You will investigate changes in various substances, like the changes described above. As well, you will learn how to identify different kinds of changes and how to explain what happens when matter changes.

The Particle Model of Matter

More than 2000 years ago in Greece, a philosopher named Democritus suggested that matter is made up of tiny particles too small to be seen. He thought that if you kept cutting a substance into smaller and smaller pieces, you would eventually come to the smallest possible particles—the building blocks of matter.

Many years later, scientists came back to Democritus' idea and added to it. The theory they developed is called the **particle model** of matter.

There are four main ideas in the particle model:

1. All matter is made up of tiny particles.

2. The particles of matter are always moving.

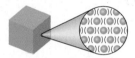

3. The particles have spaces between them.

4. Adding heat to matter makes the particles move faster.

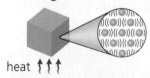

heat ↑↑↑

LEARNING TIP ◁

Are you able to explain the particle model of matter in your own words? If not, re-read the main ideas and examine the illustration that goes with each.

Scientists find the particle model useful for two reasons. First, it provides a reasonable explanation for the behaviour of matter. Second, it presents a very important idea—the particles of matter are always moving. Matter that seems perfectly motionless is not motionless at all. The air you breathe, your books, your desk, and even your body all consist of particles that are in constant motion. Thus, the particle model can be used to explain the properties of solids, liquids, and gases. It can also be used to explain what happens in changes of state (**Figure 1** on the next page).

The particles in a solid are held together strongly. The spaces between the particles are very small.

A **solid** has a fixed shape and a fixed volume because the particles can move only a little. The particles vibrate back and forth but remain in their fixed positions.

As a solid is heated, the particles vibrate faster and faster until they have enough energy to break away from their fixed positions. When this happens, the particles can move about more freely. The change from a solid to a liquid is called **melting.**

The reverse of melting is called **freezing** or solidification. This is the change from a liquid to a solid. As a liquid cools, the particles in the liquid lose energy and move more and more slowly. When they settle into fixed positions, the liquid has frozen or solidified.

The particles in a liquid are separated by spaces that are large enough to allow the particles to slide past each other.

A **liquid** takes the shape of its container because the particles can move around more freely than they can in a solid. They are held close together, however. Therefore, a liquid has a fixed volume, like a solid.

When a liquid absorbs heat energy, the particles move about more and more quickly. Some of the particles gain enough energy to break free of the other particles. When this happens, the liquid changes to a gas. The change from a liquid to a gas is called **evaporation.**

The reverse process—the change from a gas to a liquid—is called **condensation.** As a gas cools, the particles in the gas lose energy and move more and more slowly until the gas condenses to a liquid.

The particles in a gas are separated by much larger spaces than the particles in a liquid or a solid. Therefore, a gas is mostly empty space.

A **gas** always fills whatever container it is in. Since the particles are moving constantly in all directions, they spread throughout their container, no matter what volume or shape their container is.

Figure 1
Explaining changes of state using the particle model

Sublimation: A Special Change of State

Some solids can change directly to a gas without first becoming a liquid. This change of state is called **sublimation** [sub-luh-MAY-shun]. In sublimation, individual particles of a solid gain enough energy to break away completely from the other particles, forming a gas.

For example, sublimation occurs as the solid material in a room deodorizer gradually "disappears" into the air. Sublimation also occurs as a block of dry ice (frozen carbon dioxide) in an ice-cream cart "disappears" (**Figure 2**). If you live in a cold climate, you may have seen wet laundry hung outside in the winter go from frozen solid to dry because of sublimation.

Figure 2
Dry ice (frozen carbon dioxide) seems to disappear as it changes directly from a solid to a gas.

All States Have Fixed Mass

When matter changes state, it does not lose or gain mass. The mass of water vapour that is produced by melting an ice cube and then boiling the water is the same as the mass of the original ice cube.

When a liquid is poured from one container to a different-shaped container, its shape changes, but its mass does not change (**Figure 3**). If a volume of a gas is squeezed into a smaller volume, its mass does not change (**Figure 4**). We say that the mass of a specific amount of a solid, liquid, or gas is fixed.

▷ **LEARNING TIP**

Look at these photos and read the captions. Then check for understanding. Ask yourself, "What is the main idea here?"

Figure 3
Even though the shape of water changes as it is poured from one container to another, the mass of the water stays the same.

Figure 4
Gases can be squeezed into smaller containers, but the mass of the gas does not change.

Ⅲ▶ CHECK YOUR UNDERSTANDING

1. Copy **Table 1** in your notebook. Complete the table by writing "yes" or "no" in each space.

   Table 1 Summary of States

State	Fixed mass?	Fixed volume?	Fixed shape?
solid			
liquid			
gas			

2. Use diagrams and words to explain what happens to the particles of matter in each of the following situations. Are the particles moving faster or slower? Are they getting farther apart or closer together?
   a) Butter is warmed on a stove.
   b) Water vapour cools and forms raindrops.
   c) Liquid wax hardens.
   d) Water boils.
   e) Frost forms on a window.

Physical and Chemical Changes

Every day, you experience changes in matter. Cooking eggs, burning leaves, freezing water, and mixing oil and vinegar to make salad dressing involve changes in matter. Understanding and categorizing these changes are an important first step in learning how to use them.

TRY THIS: BRAINSTORM CHANGES

Skills Focus: observing, communicating, recording

1. In a small group, brainstorm a list of changes in matter. Use a different action word for each change. For example, concrete *hardens*, wood *rots*, snow *melts*, paper *yellows*, and fireworks *explode*.

2. Which changes do you think result in a new substance being formed? Indicate these with a check mark (✔).

3. Which changes do you think add materials to the air? Indicate these changes with an asterisk (*).

Brainstorm Changes	
wood burns	*

Physical Changes

In a **physical change,** the substance that is involved remains the same, even though its form or state may change. A piece of wood cut into pieces is still wood (**Figure 1(a)**). When ice melts, it is still ice (**Figure 1(b)**). Changes of state—melting, freezing, evaporation, condensation, and sublimation—are physical changes.

a) Sawing wood

b) Melting ice

Figure 1
Physical changes

In a physical change, the particles of a substance may move closer together or farther apart, or they may mix with particles of other substances. However, no new kinds of particles are produced. Dissolving is a physical change. When you dissolve sugar in water, the sugar particles spread out and mix with the water particles, but they are still there. You can reverse the process by evaporating the water and collecting the sugar.

Changes that can be reversed are called **reversible changes.** Physical changes are often reversible, but not always. You can reverse the physical change that occurs when you melt ice by cooling the water until it freezes again. You cannot reverse the physical change that occurs when wood is sawed into pieces. Changes that cannot be reversed are called **non-reversible changes.**

Chemical Changes

In a **chemical change,** the original substance is changed into one or more different substances with different properties. When a candle burns, it becomes shorter. Some wax may melt down the side of the candle, but some seems to disappear. Where does the wax go? As the wax burns, some wax particles react with oxygen in the air to produce water vapour, carbon dioxide gas, heat, and light. The wax particles that seem to disappear are actually changing into other substances.

Burning a log and frying an egg are also chemical changes (**Figure 2**). When you fry an egg, the liquid egg white part of the egg changes colour and becomes solid. The cooked egg has properties that are different from the properties of the uncooked egg. When you burn a log, you can see it getting smaller. You can feel the heat and see the light given off. You can also see new materials, such as ash and smoke.

Chemical changes always involve the production of new substances. Most chemical changes are difficult to reverse.

Figure 2
Chemical changes

a) Burning wood

b) Cooking eggs

The Importance of Chemical Changes

You rely on chemical changes to survive. The clothes you wear and the food you eat are the results of chemical changes. There are millions of chemical changes going on around you. Some are even happening in your body. Plants use energy from the Sun to combine water and carbon dioxide, which react to form sugar and oxygen. When you eat these plants and inhale oxygen from the air, the sugar and oxygen react in your cells to produce water, carbon dioxide, and energy. You need the energy from this reaction for your daily activities.

Determining Whether a Change Is Physical or Chemical

You cannot see the chemical change in wax by looking at a burning candle. You can often see the results of a chemical change, however. You can see the light from a candle, and the colour and firmness of a cooked egg. So, how can you tell if a chemical change has occurred? How can you tell the difference between a chemical change and a physical change? **Figure 3** shows five clues that a chemical change has occurred.

LEARNING TIP ◁

Make notes on evidence of chemical change in a five-column chart. Copy the illustrations and captions from **Figure 3** as the column headings. Add examples under each heading.

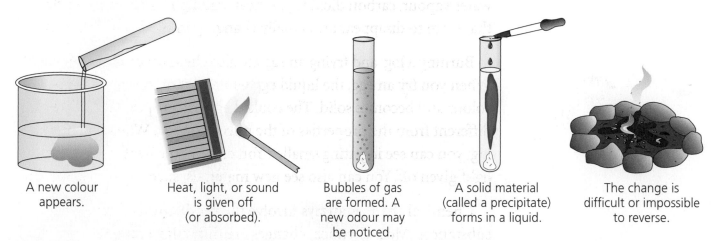

| A new colour appears. | Heat, light, or sound is given off (or absorbed). | Bubbles of gas are formed. A new odour may be noticed. | A solid material (called a precipitate) forms in a liquid. | The change is difficult or impossible to reverse. |

Figure 3
Evidence of a chemical change

When classifying changes, do not jump to conclusions too quickly. The clues in **Figure 3** suggest that a new substance has been produced, but any one of them could also accompany a physical change. You must consider several clues in order to determine what type of change has taken place.

Skills Focus: observing, classifying

Look at each of the following photos. Classify the change that is shown as physical or chemical, and reversible or non-reversible.

▬▶ CHECK YOUR UNDERSTANDING

1. Explain how a physical change differs from a chemical change. Present your explanation in a table.

2. Give three examples of reversible physical changes that were not mentioned in this section. Give one example of a non-reversible physical change that was not mentioned in this section.

3. What clues might you observe if a chemical change is occurring?

4. In an experiment, Ethan and Deepa tested different combinations of substances. They made the following conclusions. Are their conclusions valid? Explain your reasoning.
   a) When we opened a can of cola, it fizzed. This showed that a chemical change had occurred.
   b) When Ethan's Dad sawed through a piece of wood, smoke came up around the blade of the power saw. The sawdust was blackened around the edges of the blade. A chemical change had occurred, because the sawdust appeared different from the original wood.
   c) Heat and light were given off by a light bulb. A chemical change took place in the light bulb.

5. When you mix sugar in water, the sugar crystals disappear. Explain why this is an example of a physical change rather than a chemical change.

Fireworks—Extreme Chemical Changes

Have you ever been to a fireworks display and wondered how the colours and sounds are produced? They are produced by chemical changes in the different substances in the fireworks. Each firework is a carefully controlled series of chemical changes, which occur at just the right times and produce large amounts of heat.

A typical firework contains a fuel, a source of oxygen, a fuse (a source of heat to set off the reaction), and a colour producer. Suppose that a firework is expected to rise 50 m and then produce a red burst of fire, followed by a bright flash. This firework would have to contain three different combinations of substances to produce three different chemical changes: one to lift the firework and two to create the two explosions.

The main part of a firework is the fuel and the source of oxygen. When these react with substances such as aluminum or sulfur, a large amount of heat, a loud bang, and flashes of light are produced.

Different substances produce different effects. For example, iron filings and charcoal (carbon) produce gold sparks. Strontium carbonate produces a red flame. Potassium benzoate produces a whistling sound. **Table 1** summarizes some of the other chemicals involved in fireworks. The next time you go to a fireworks display, think about the chemistry involved!

Table 1 Some Chemicals Used for Special Effects

Materials	Special Effect
magnesium metal	white flame
sodium oxalate	yellow flame
barium chlorate	green flame
potassium nitrate and sulfur	white smoke
potassium perchlorate, sulfur, and aluminum	flash and bang

● SKILLS MENU

○ Questioning	● Observing
○ Predicting	○ Measuring
○ Hypothesizing	● Classifying
○ Designing Experiments	● Inferring
○ Controlling Variables	● Interpreting Data
○ Creating Models	● Communicating

Name the Change

You have learned that there are two types of changes: physical changes and chemical changes. It is not always easy to tell the difference between a physical change and a chemical change. Each clue must be carefully interpreted. In this investigation, you will combine samples of familiar matter. Then you will decide if a physical change or a chemical change has occurred.

Question

Can you tell the difference between a physical change and a chemical change?

Materials

safety goggles
apron
beakers
vinegar
medicine dropper
water
baking soda
milk
egg shell
uncooked spaghetti
lemon juice
paper
oven mitts
hot plate
yeast
sugar

- safety goggles
- apron
- small beakers or small glass jars
- medicine dropper
- vinegar
- water
- baking soda
- milk
- small piece of eggshell
- 2 pieces of uncooked spaghetti

- lemon juice
- paper
- oven mitts
- hot plate
- yeast
- sugar

> ✋ **Wear safety goggles and an apron.**

▶ Procedure

1 Make a table like the one below to record your observations.

2 Observe and record the properties of the substances before you combine them.

Data Table for Investigation 5.3

Procedure	Observations before	Observations after	Physical change or chemical change?
vinegar and baking soda	Vinegar is clear and colourless, and smells sharp. Baking soda is ...		
water and baking soda			

3 Using a medicine dropper,
- add vinegar to a small sample of baking soda
- add water to a small sample of baking soda
- add vinegar to a small sample of milk
- add water to a small sample of milk

- add vinegar to a small piece of eggshell
- add water to a small piece of eggshell

Observe and record your results and whether you saw a physical or chemical change.

 Use non-flammable oven mitts for step 4. Do not let the paper touch the hot plate.

4 Dip a piece of uncooked spaghetti in water. Use the spaghetti like a pen to write your initials on a piece of paper. Dip another piece of uncooked spaghetti in lemon juice and write your initials on another piece of paper. Put on oven mitts, and heat both papers gently over a hot plate. Observe and record your results.

5 Mix yeast and a small amount of warm water in two containers. Stir some sugar into one of the containers. Observe and record your results.

Analyze and Evaluate

1. Which combinations produced physical changes? Which combinations produced chemical changes? What clues did you use to decide?

2. Is appearance a good clue to the type of change that has occurred? Why or why not?

Apply and Extend

3. Based on your observations, why do you think recipes call for baking soda?

4. Give one example of a physical change and one example of a chemical change that might occur when preparing a meal.

⫸ CHECK YOUR UNDERSTANDING

1. Why did you add both water and vinegar to the baking soda, milk, and eggshell? Why did you write with both water and lemon juice on the paper?

LEARNING TIP ◁

For a review of variables, see "Controlling Variables" in the Skills Handbook.

Chemical Changes in the Environment

Changes are constantly occurring in the environment. Matter may become part of the atmosphere, sit in a landfill, be washed away to an ocean, or be buried underground. However, matter is never completely gone. It remains on Earth. Matter can turn into something else and be used again and again because of chemical changes. One change is followed by another and another.

Chemical Changes in the Living Environment

There are many examples of chemical changes in the living environment. One of the most spectacular examples is a forest fire. A forest fire is not only the end of a forest; it is also the beginning of a new forest (**Figure 1**). In a forest fire, huge trees seem to disappear in minutes. They have not actually disappeared, however. The materials in the trees have been changed into other materials. The leaves and trunks have become gases and smoke in the air, and ashes on the ground. How do you think these new materials can be used as new growth begins?

Figure 1
Whole forests can be consumed by fire, which is a chemical reaction. Chemical reactions are also involved in the gradual regrowth of the forest.

Not all chemical changes are as spectacular as burning. Many are so slow that you cannot see them happening. For example, the new growth in a burned-over area is the result of many chemical changes that go on inside living organisms. Similarly, when the trees in a forest die and decay, chemical changes slowly return the matter in the trees to the environment.

Chemical Changes in the Non-Living Environment

Many chemical changes that do not involve living things also occur in nature.

One very common chemical change is what happens to metals that contain iron, especially when they are wet. This change is called rusting (**Figure 2**). You can see the product of this change—rust—on old bicycles, metal gardening tools, and old cars that have been through many seasons of rain and snow. The rust is soft and flaky—very different from the original shiny metal. When iron rusts, it combines with oxygen in the air to form a new substance.

LEARNING TIP ◁

Compare this information with what you already knew about rusting. Ask yourself, "Is there any information here that is new to me?"

Figure 2
Iron reacting with oxygen to produce rust is an example of a chemical change. Rust damages objects made of metal, such as bicycles and cars.

Figure 3
When silver reacts with oxygen, the silver turns black. Silver tarnishing is a chemical change.

Figure 4
When copper tarnishes, it turns green, like the roof of Hotel Vancouver.

Other metals, such as silver and copper, also combine with oxygen in the air. The new substances that are formed are a different colour than the original silver and copper (**Figures 3** and **4**).

Many industries carry out chemical changes to make the materials that you use every day. Plastics (including vinyl and polyester) are all products of chemical changes. In the mining industry, chemical changes are carried out to separate valuable metals from rock.

⫸ CHECK YOUR UNDERSTANDING

1. How is a forest fire an example of a chemical change? List specific clues that support your answer.

2. What is rusting?

3. Name three chemical changes that do not require living organisms.

Materials Scientists

Materials scientists research the structures and chemical properties of various materials in order to develop new materials, or enhance existing materials to fit new applications.

Figure 1

Many materials scientists work in research and development (R&D). In basic research, they investigate the properties, structure, and composition of matter and how elements and compounds react to each other. In applied R&D, they use the knowledge from research to create new products and processes, or improve existing ones.

Chemistry plays a big role in materials science, because it provides information about the properties, structure, and composition of matter. But materials science covers a broad range of sciences. For example, materials scientists have worked with medical professionals to develop materials that can be used to repair and replace body parts. Together they have developed artificial joints, heart valves, ears, and even cochlear implants that allow deaf people to hear.

Sometimes materials scientists find diverse uses for similar materials. The Teflon used to coat non-stick frying pans, the polyester Dacron used in clothing, and the Gore-Tex used for rain jackets are all also used to make artificial blood vessels.

Space exploration depends on materials scientists developing materials that can withstand the temperature extremes, radiation, and other hazards of space (**Figure 1**).

Research in materials science has led to improvements in common materials as well. Coatings and paints (**Figure 2**) that resist corrosion have been developed. However, most paints that resist corrosion contain chromate, a very poisonous chemical that can

Figure 2

pollute water supplies. Materials scientists are working on a new corrosion resistant paint with a "smart" pigment that absorbs corrosion-causing chemicals, and releases a safer corrosion inhibitor that forms a protective film over cracks in the paint.

Some of the most exciting materials science research is in the field of electronics. Materials scientists have found ways to dramatically reduce the size of integrated circuit chips (**Figure 3**), allowing for smaller and smaller electronic goods.

Figure 3

From non-stick bandages to nanotechnology, it is a very exciting time to have a career in materials science.

Chapter Review

Matter is made up of moving particles.

Key Idea: The behaviour of matter can be explained using the particle model.

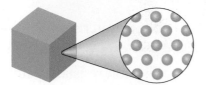

All matter is made up of tiny particles.

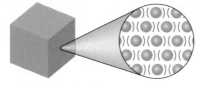

The particles of matter are always moving.

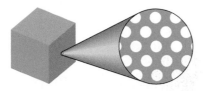

The particles have spaces between them.

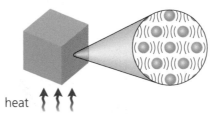

heat ↑ ↑ ↑

Adding heat to matter makes the particles move faster.

Vocabulary

particle model p. 117

solid p. 118

melting p. 118

freezing p. 118

liquid p. 118

evaporation p. 118

condensation p. 118

gas p. 118

sublimation p. 119

Key Idea: Matter can undergo physical and chemical changes.

Physical change

Chemical change

Vocabulary

physical change p. 121

chemical change p. 122

Key Idea: Changes in matter can be reversible or non-reversible.

Vocabulary

reversible changes p. 122

non-reversible changes p. 122

Reversible change

Non-reversible change

Key Idea: Chemical changes can be distinguished by observable clues.

A new colour appears.

Bubbles of gas are produced.

Heat, light, or sound is produced or absorbed.

A new material is produced.

The change is difficult or impossible to reverse.

Key Idea: Chemical changes occur in our living and non-living environments.

Review Key Ideas and Vocabulary

When answering the questions, remember to use vocabulary from the chapter.

1. Copy the following table into your notebook, and complete it.

State	solid	liquid	gas
Type of particle movement	back and forth		
Spaces between particles		wider	

2. Using the particle model, explain what happens to water as it is gradually heated and changes from ice to steam.

3. Give an example of a physical change that is reversible and a physical change that is not reversible.

4. Suggest five clues that you would consider before deciding whether a change is a physical change or a chemical change.

5. State whether each change is a physical change or a chemical change. Give at least one reason for your answer.
   a) Frost forms on windows.
   b) Tea is made using hot water and a tea bag.
   c) A firecracker explodes.
   d) Concrete becomes hard after it is poured.
   e) The burner on an electric stove glows red.
   f) Coffee changes colour when cream is added.
   g) Liquid nitrogen boils at −196°C.
   h) Butter is heated in a frying pan until it turns brown.

   i) When a flame is brought near hydrogen gas in a test tube, there is a loud pop.

6. Give an example of a chemical change that occurs in your living environment and a chemical change that occurs in your non-living environment.

Use What You've Learned

7. Use the particle model of matter to explain why it is easier to move your hand through air than through water.

8. Solids are described as having a fixed volume. Most solids expand (increase volume) slightly when heated, however.
   a) Use the particle model of matter to explain this observation.
   b) Many bridges have expansion joints in them (**Figure 1**). Research expansion joints. Determine what would happen to a bridge on a hot day if it did not have expansion joints.

www.science.nelson.com

Figure 1
Why do bridges have expansion joints?

9. Microwave ovens cook food quickly (**Figure 2**). Research how microwave ovens work. Use the particle model to explain why they cook food so quickly.

www.science.nelson.com **GO**

Figure 2
How do microwave ovens work?

10. Give examples of physical and chemical changes that are useful to you. Think of a way to display your examples.

11. What affects how quickly a certain type of metal rusts (**Figure 3**)? Design an experiment, using iron nails, to test what speeds up this chemical change. Put each nail in a separate jar. Add water, lemon juice, vinegar, salt water, and other liquids. Observe what conditions make the nail rust more quickly. Make a poster to illustrate your experiment and your results.

Figure 3
What would make these nails rust quickly?

Think Critically

12. Which of the changes described in **Figure 4** involve a chemical change? Which involve a physical change? Determine the impact each change could have on the environment. Which changes must car designers consider in order to minimize damage to the environment?

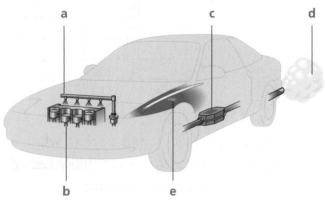

Figure 4
a) In the fuel injector on top of the engine, liquid gasoline is evaporated and mixed with air.
b) Inside the engine cylinders, the gasoline burns very rapidly, producing hot exhaust gases, including water vapour, carbon dioxide, and nitrogen oxides.
c) The exhaust gases pass through the catalytic converter, where some harmful gases are changed into different gases that are less harmful to our environment.
d) The exhaust passes out the tailpipe. On a cold day, steam from the exhaust condenses into a white cloud.
e) As the steel of the car is exposed to air and water, a crumbly reddish-brown substance forms: the steel has changed into rust.

13. Are physical changes in matter or chemical changes in matter more important to your life? Explain your answer.

Reflect on Your Learning

14. How has learning about the particle model changed how you think about matter in your environment?

CHAPTER 6

Matter can be classified.

Matter is made up of many different substances. Some of these substances are similar to each other, and some are very different. You can see and feel most substances. You can also describe them, based on what you see and feel—for example, gold is solid, hard, and shiny.

The properties of matter that you studied in Chapter 4 can be used to classify matter. Classifying matter helps you predict properties of similar substances. For example, silver has many of the same properties as gold and, like gold, is used to make jewellery. Classifying matter also helps you predict how substances will behave when they are mixed with other substances. The gold that is used in jewellery is not pure gold. It is mixed with other metals to make it stronger. In this chapter, you will learn some of the ways that scientists classify matter.

Pure Substances and Mixtures

As you have already learned, all matter is made up of particles. There are many different kinds of particles. Different substances have different properties because they contain different kinds of particles.

Pure Substances

A substance that contains only one kind of particle is called a **pure substance.** There are millions of pure substances, but only a few can be found in nature. For example, water is a pure substance, but pure water is difficult to find in nature. Even the clearest spring water contains dissolved minerals. In nature, pure substances tend to mix with other substances. Diamonds are one of the few exceptions (**Figure 1**). They are formed deep within Earth, in only a very few areas.

Figure 1
A diamond is an example of a pure substance. All the particles in a diamond are the same.

Most of the pure substances that you encounter in your daily life have been made pure by people through refining. Aluminum foil is a pure substance, and so is table sugar (**Figure 2**).

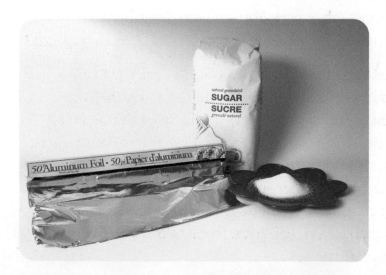

Figure 2
Aluminum foil and sugar are pure substances.

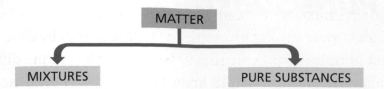

All samples of a pure substance have the same properties, no matter what size the samples are or where in the world the samples are found. For instance, all samples of pure gold have the same melting and boiling points and the same density. Because every sample of a pure substance has the same properties, scientists have made reference lists of pure substances and their properties. These reference lists can help you to identify an unknown substance based on its properties.

Mixtures

Almost all the natural substances and manufactured products in the world are mixtures of pure substances. A **mixture** contains two or more pure substances, as shown in **Figure 3**.

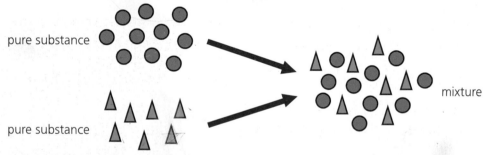

Figure 3
Most substances that you encounter are mixtures. Mixtures contain at least two pure substances.

Mixtures can be any combination of solids, liquids, and gases. For example, soft drinks are mixtures made from liquid water, solid sugar, and carbon dioxide gas (**Figure 4**).

Figure 4
Soft drinks may look like pure substances, but they are mixtures.

Breads are mixtures of yeast, flour, sugar, water, air, and other chemicals (**Figure 5**). The properties of mixtures may be different in different samples because there may be more or less of the different kinds of particles. For example, breads do not always have the same number of yeast or sugar particles in them.

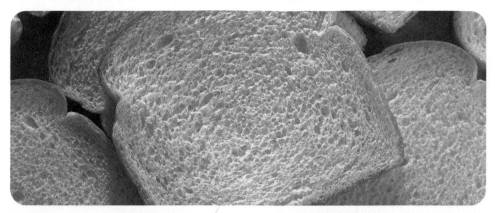

Figure 5
Bread is a mixture of different substances.

TRY THIS: *TEST INK*

Skills Focus: observing, interpreting data

1. Cut a "tongue" in a piece of filter paper.

2. About 1 cm from the end of the tongue, draw a large dot with a black water-soluble marker.

3. Put the filter paper on top of a 250-mL beaker, with the tongue bent down into the beaker.

4. Carefully add water until it touches the filter paper tongue but does not touch the dot. Observe what happens as the water soaks into the filter paper. Is ink a pure substance or a mixture?

▐▶ CHECK YOUR UNDERSTANDING

1. Explain the difference between a pure substance and a mixture, using examples of each.

2. Explain the difference between a pure substance and a mixture, using the particle model.

3. Give three examples of pure substances and three examples of mixtures.

6.1 Pure Substances and Mixtures

6.2 *Elements and Compounds*

There are millions of pure substances. Can anyone expect to learn about all of them? How would you start? How would you find out which ones are safe? How would you find out which ones are useful?

People have investigated pure substances for thousands of years. Ten thousand years ago, people learned how to extract copper from rocks by heating the ore. In medieval times, alchemists [AL-ku-mists] tried to break down metals, such as copper, to make gold. They dissolved and mixed various substances, filtered, and heated. None of the alchemists ever succeeded in making gold. They discovered, however, that some pure substances can be broken down into other pure substances, while others cannot. For this reason, pure substances are classified into two types: elements and compounds.

Elements

Elements are pure substances that cannot be broken down into any other pure substances. After many investigations, scientists found that there are only about 104 pure substances that are elements.

Elements are composed of only one kind of particle. For instance, aluminum foil is made of the element aluminum. It is composed of only particles of aluminum (**Figure 1**).

Figure 1
The element aluminum in aluminum foil is composed of aluminum particles.

Some elements, such as iron, aluminum, and oxygen, are common in nature, although they are usually found mixed with other substances. Other elements, such as krypton, are extremely rare. Some elements are considered safe. Other elements, such as sodium and chlorine, are explosive or poisonous.

Compounds

Elements can combine with other elements to form new pure substances, called compounds. **Compounds** are pure substances that are made up of two or more different elements. Compounds are related to elements in the same way that words are related to the letters of the alphabet. Thousands of words can be made from the 26 letters of the English alphabet. Similarly, millions of compounds can be made by combining the 104 elements.

Compounds can be solids, liquids, or gases. One example of a compound is water. Water is made up of the elements hydrogen and oxygen (**Figure 2**). Thus, a particle of water contains both hydrogen and oxygen. Every particle of water is the same as every other particle of water. At one time, scientists thought that water was made up of particles that could not be broken down further. Scientists now know, however, that water can be broken down into hydrogen and oxygen.

H_2O

Figure 2
Water is a compound composed of hydrogen and oxygen particles.

The elements in some common compounds are listed in **Table 1**.

Table 1 Elements in Some Common Compounds

Compound	Elements combined in compound
water	hydrogen and oxygen
table salt (sodium chloride)	sodium and chlorine
carbon dioxide	carbon and oxygen
sugar (any type)	carbon, hydrogen, and oxygen
alcohol (any type)	carbon, hydrogen, and oxygen
chalk (calcium carbonate)	calcium, carbon, and oxygen
baking soda	sodium, hydrogen, carbon, and oxygen

Different elements have different properties because they have different particles. In the same way, different compounds have different properties because they have different combinations of elements. The properties of a compound can be very different from the properties of the elements that make it up. Table salt (**Figure 3**) is made of two elements, called sodium and

Figure 3
Table salt

chlorine. Sodium on its own is a soft, silvery metal that is poisonous and reacts violently with water (**Figure 4**). Chlorine is a greenish-yellow gas that is extremely poisonous (**Figure 5**). Each of these elements could be fatal if consumed on its own—for example, if you breathed in too much chlorine or swallowed a large quantity of sodium. When sodium and chlorine combine, however, they form table salt (sodium chloride), which you can safely eat and need in your diet.

Figure 4
Sodium metal

Figure 5
Chlorine gas

▷ **LEARNING TIP**

Go back to the graphic organizer you started in section 6.1. Add "elements" and "compounds" under "pure substances." Your graphic organizer should now look like this: ▶

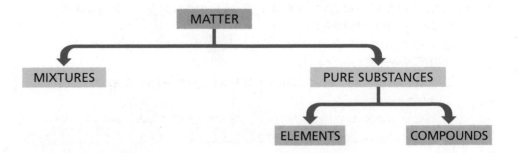

TRY THIS: CLASSIFY MODELS OF MATTER

Skills Focus: creating models, classifying

Copy your graphic organizer onto a large piece of paper.
Your teacher will give you eight jars, containing the following items
(**Figure 6**):

1. five nuts

2. five bolts, five nuts, and five washers

3. five bolts with nuts attached

4. five bolts with a nut attached and five bolts with a washer and a nut attached

5. five bolts

6. five nuts and five washers

7. five bolts with a washer and a nut attached

8. five washers

Figure 6

Each jar is a model, representing a different type of matter. Each bolt, nut, and washer represents a different type of particle. Classify the eight models of matter as elements, compounds, or mixtures by placing them in the appropriate places on your graphic organizer.

LEARNING TIP ◁

If you are having difficulty remembering the differences between mixtures, pure substances, elements, and compounds, scan the text for the information you need and make notes on your graphic organizer before you try to classify the models.

▶ CHECK YOUR UNDERSTANDING

1. Explain the difference between an element and a compound, using examples of each.

2. Explain the difference between an element and a compound, using the particle model.

3. State whether each pure substance is an element or a compound. Explain your reasoning.
   a) a clear, colourless liquid that can be split into two gases with different properties
   b) a yellow solid that always has the same properties and cannot be broken down
   c) a colourless gas that burns to produce carbon dioxide and water

Mixtures

Most of the substances you use in your daily life are not pure substances. For example, hand lotion, shampoo, and soap are made of many substances, such as colouring and perfumes, mixed together. Foods contain preservatives and other additives. Even fruit juice that is labelled "100% pure" is actually a mixture of water, citric acid, and other substances (**Figure 1**).

Figure 1
This "pure" apple juice is a mixture.

Classifying Mixtures

If you were asked to name some pure substances, you might think of common substances such as sugar, water, salt, and oxygen gas. Other substances you might think of may seem to be pure, even though they are not. For example, how would you classify vinegar? Is it a pure substance or a mixture? To be able to classify matter, you need to know more about mixtures. One way that scientists classify mixtures is to group them according to their appearance.

▷ **LEARNING TIP**

Before you read further, look at the subheadings on the next two pages. Predict how many categories scientists use to classify mixtures.

Mechanical Mixtures

A **mechanical mixture** is a mixture in which two or more different parts can be seen with the unaided eye. Granola cereal is an example of a mechanical mixture (**Figure 2**). Concrete is another example.

Figure 2
This cereal is a mechanical mixture. What other foods can you classify as mechanical mixtures?

Suspensions

A **suspension** is a cloudy mixture in which clumps of a solid or droplets of a liquid are scattered throughout a liquid or gas. Muddy water and tomato juice are suspensions. The parts of a suspension may separate into layers if the suspension is not stirred.

Farm-fresh milk is a suspension. If the milk is left standing, the fatty part (the cream) floats to the top and the watery part sinks to the bottom (**Figure 3**). Commercially available milk does not separate. It is a special kind of suspension, called an **emulsion,** which has been treated to keep it from separating. In a process called homogenization, the milk is sprayed through very small openings. This breaks down the fat into droplets that are so tiny they stay suspended.

Figure 3
Cream floats to the top of farm-fresh milk.

Solutions

A **solution** is a mixture that appears to be only one substance. The parts of a solution are so completely mixed that they cannot be seen, even under a microscope. This is because the particles of the substance that dissolves fill in the spaces between the particles of the substance it dissolves in. Clear apple juice (a liquid) (**Figure 4**), clean air (a gas), and stainless steel (a solid mixture of metals) are all solutions.

Figure 4
Apple juice is a solution.

▷ **LEARNING TIP**

For a review on models, see "Creating Models" in the Skills Handbook.

TRY THIS: *MODEL A SOLUTION*

Skills Focus: modelling, predicting, observing

You can make a model to show how particles mix in a solution. The advantage of making a model is that you can observe a process you would not normally be able to see.

1. Half fill a clear plastic container with marbles. On the outside of the container, mark the level of the marbles with a marker. Then half fill a second, identical container with sand.

2. Predict the total volume that will result when you combine the marbles and the sand by marking the level you think will result.

3. Carefully pour the sand into the container with the marbles, and shake gently. How accurate was your prediction of the total volume? Explain.

4. How is the container of sand and marbles like a solution? How is it different?

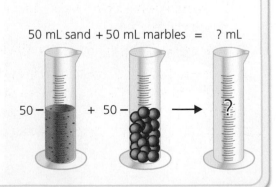

50 mL sand + 50 mL marbles = ? mL

50 —   + 50 —   →   ?

Examples of Mixtures

Table 1 gives examples of mechanical mixtures, suspensions, and solutions. Can you explain the classification of each substance listed?

Table 1 Examples of Mechanical Mixtures, Suspensions, and Solutions

Mechanical mixtures	Suspensions	Solutions
snow falling through the air	foggy air	clean air
salad	salad dressing	vinegar
cornflakes and milk	orange juice	tea
concrete (cement, sand, and gravel)	muddy water	tap water
abrasive skin cleanser	hand lotion	clear shampoo

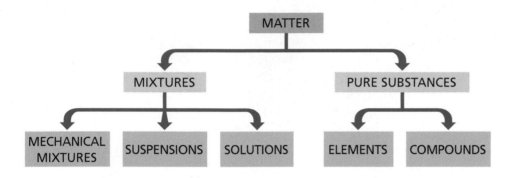

LEARNING TIP

Go back to the graphic organizer you started in section 6.1. Complete it by adding "mechanical mixtures," "suspensions," and "solutions" under "mixtures." Your graphic organizer should now look like the one on the left.

▶ CHECK YOUR UNDERSTANDING

1. List at least three mechanical mixtures and three solutions from your everyday life that were not mentioned in this section.

2. State whether each substance is a mechanical mixture, a suspension, or a solution. Explain your reasoning.
   a) green relish
   b) freshly squeezed orange juice
   c) soda pop in a glass
   d) bubble tea
   e) trail mix
   f) traditional Aboriginal paint, made of red ochre and grease
   g) vegetable soup

3. How are suspensions and solutions similar? How are they different?

4. Suppose that you dissolve 250 mL of drink crystals in 1000 mL of water. You get 1175 mL of drink rather than 1250 mL. How can you use the particle model of matter to explain this?

6.4 Separating Mixtures

▷ **LEARNING TIP**

Before reading this section, "walk" through it, looking at the headings. What ways of separating mixtures do you think you will learn about?

Does your family have a "junk drawer" somewhere, maybe in the kitchen or near the door (**Figure 1**)? Have you ever tried to sort out all the items that have collected in the drawer? In everyday life, there are many situations in which people want to separate the parts of a mixture. For example, you do not want to drink water that contains algae or fish, or dissolved chemicals from factories. You prefer to have these removed from the water before it is pumped to your home. Harmful or toxic substances from factories must be removed from any waste products before the waste products can be released into the environment.

Figure 1
How could the boy separate the items in the drawer?

Depending on the mixture involved, separating the parts can be easy or difficult. In this section, you will learn about some ways to separate different types of mixtures.

Picking Apart

Figure 2
Picking apart a mixture

You would probably separate the mixture in a junk drawer by simply taking out the different items—tools, elastic bands, scrap paper, and so on. You would use observable properties, such as shape and colour, to separate the mixture. If the pieces in a mixture are smaller, you might have to use a magnifier and forceps. Picking apart works when you can easily see the different pieces (**Figure 2**). It only works well for small quantities of mixtures. It is too time-consuming to use for large quantities.

Filtering

You can remove solid pieces of matter from a liquid or gas by passing the mixture through a device that allows smaller particles to pass through but holds back larger particles. This is called filtering (**Figure 3**). Drinking water is an example of a mixture that is filtered. The water passes through a filter, which allows the liquid through but holds back larger particles. The liquid that passes through is called the filtrate and the solid material that is held back by the filter is called the residue.

There are many other examples of filtration. Air is filtered in car engines and factory smokestacks. Window screens act as filters to keep flies and mosquitoes out of homes. Workers who use spray paint wear facemasks so that they do not breathe in droplets of paint. Tea bags keep tea leaves out of tea, and coffee filters keep coffee grounds out of coffee.

Even very small pieces of substances can be removed from mixtures by filtration if the holes in the filter are small enough. Thus, filters can be used to separate solids from mechanical mixtures or suspensions. Filters cannot be used, however, to separate parts of solutions.

Figure 3
Filtering a mixture

Using Density

Density can also be used to separate mixtures. If the substances in a mechanical mixture have different densities, one substance may float and another may settle to the bottom (**Figure 4**). For example, density can be used to separate a mixture of sand and wood chips. If water is added to the mixture, the wood chips float and the sand sinks, making the mixture easy to separate.

Figure 4
Using density to separate a mixture

Using Magnetism

Magnetism can be used to separate a mechanical mixture if one of the substances in the mixture is attracted to a magnet (**Figure 5**). This works well for a mixture of iron filings in sand.

iron filings and sand

Figure 5
Using magnetism to separate a mixture

6.4 Separating Mixtures

Dissolving

You can sometimes **dissolve** one of the substances in a mixture. When a substance dissolves, it mixes completely with another substance. For example, if you add water to a mixture of sand and salt, the salt dissolves. This makes the undissolved sand easier to separate out of the mixture by filtering (**Figure 6**).

Figure 6
Dissolving one of the substances in a mixture

salt and sand

Evaporating

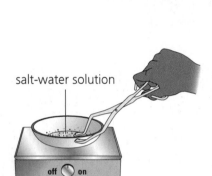

You can also evaporate part of a solution to get a substance dissolved in it (**Figure 7**). For example, you can evaporate the water from a cup of tea. The solid that remains is the tea. Sometimes, the solid that remains crystallizes. For example, when the water evaporates from a salt-water solution, the salt crystallizes.

salt-water solution

Figure 7
Evaporating one of the substances in a mixture

⫸ CHECK YOUR UNDERSTANDING

1. Describe the method you would use to separate each mixture in **Figure 8**.

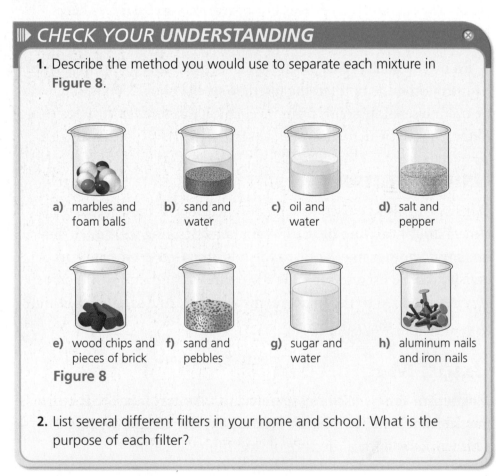

a) marbles and foam balls

b) sand and water

c) oil and water

d) salt and pepper

e) wood chips and pieces of brick

f) sand and pebbles

g) sugar and water

h) aluminum nails and iron nails

Figure 8

2. List several different filters in your home and school. What is the purpose of each filter?

Separating a Mystery Mixture

Often, scientists do not know exactly what substances are in a mixture before they try to separate it. Sometimes they have to separate something out of a mixture to use in a different test. In this activity, you will be provided with a mystery mixture. You must determine what the different substances in the mixture are.

LEARNING TIP ◁

For more information on the steps in problem solving, see the "Solving a Problem" in the Skills Handbook.

Problem

For this activity, your teacher will give you a mystery mixture to separate.

Task

Your task is to separate the substances in the mixture using methods you learned in section 6.4. You should be able to identify six different substances.

Criteria

To be successful, your procedure must

- allow you to separate all six of the substances in the mixture
- be clear enough for someone else to follow and get the same results

Plan and Test

1. Look at your mixture. Are there any easily observable properties that give you clues about how to proceed? What separation methods can you use? In what order will you use them?

2. Design a procedure to separate your mystery mixture. Remember the different ways to separate a mixture: picking apart, filtering, using density, using magnetism, dissolving, and evaporating.

3. Decide what materials you will need. Draw a diagram to show how you will set up the equipment. Your diagram should be at least half a page in size.

4. Submit your list of materials, diagram, and procedure to your teacher for approval. Your procedure must include any safety precautions and an observation table.

5. Carry out your procedure.

Evaluate

6. Were you able to separate your mystery mixture into six separate substances? What substances did you find in your mixture?

Communicate

7. Draw a flow chart to show how you separated your mystery mixture. Post your flow chart, and compare it with the flow charts that your classmates have drawn.

▷ **LEARNING TIP**

For more information on how to create a flow chart, see "Using Graphic Organizers" in the Skills Handbook.

▐▶ CHECK YOUR UNDERSTANDING ⊗

1. Did any of the methods you tried fail to separate a substance from your mixture? If so, why?

2. What physical properties did you use to separate each substance from your mixture?

3. Are there any other methods you could have used to separate your mixture?

4. Do you think you recovered all of each substance in your mixture? How might you improve your procedure to recover as much as possible of each substance?

Concentration

When a solid dissolves in a liquid, the liquid that does the dissolving is called the solvent. The solid that dissolves is called the solute. For example, in a solution of orange-drink crystals and water, the water is the solvent and the orange-drink crystals are the solute (**Figure 1**).

Figure 1
The drink on the right has more solute than the drink on the left. How can you tell?

Have you ever made a drink by dissolving drink crystals in water and found that it tasted watery? This happens when you do not have the right concentration of solute in the solvent. **Concentration** [kon-suhn-TRAY-shun] is the amount of solute that is dissolved in a given quantity of solvent or solution.

Solutions that are made with the same substances may contain different amounts of each substance. A solution with a low concentration of solute is said to be **dilute** [die-LOOT]. A solution with a higher concentration of solute is said to be more concentrated. For example, lemonade with a small amount of dissolved sugar is a more dilute solution than lemonade with a lot of dissolved sugar. The lemonade with more sugar is a more concentrated solution. It tastes sweeter than the more dilute solution.

LEARNING TIP ◁

The key vocabulary words on this page are illustrated. If you are having trouble with these vocabulary words, look at **Figure 1** for clarification.

▷ **LEARNING TIP**

Make a web to show what you already know about substances that dissolve in water.

Solubility

You can make orange drink because the orange-drink crystals dissolve in water. Another way to say that a substance dissolves in water is to say that it is soluble in water. Can you think of some other substances that are soluble in water? If the orange-drink crystals did not dissolve in water, you would not be able to make the drink. An insoluble substance is a substance that does not dissolve. Can you think of some substances that are insoluble in water? Can a substance that is insoluble in water be dissolved in another solvent?

TRY THIS: OBSERVE DIFFERENT SOLVENTS

Skills Focus: predicting, observing, classifying

Predict whether salt, sugar, butter, and wax will dissolve in water and in ethanol. Now try to dissolve each of these solutes in the two different solvents (**Figure 2**). Record your results.

Observe Different Solvents		
	Dissolves in water?	Dissolves in ethanol?
salt		
sugar		
butter		
wax		

Figure 2
Adding sugar to ethanol to see if the sugar dissolves

Saturated and Unsaturated Solutions

Even if a substance is soluble in a solvent, there is usually a limit to how concentrated the solution can become. For example, there is a limit to how many orange-drink crystals you can dissolve in a glass of water. Imagine that you add more and more drink crystals to a glass of water, stirring constantly. Eventually, the drink crystals will just stay at the bottom of your glass (**Figure 3**). The drink solution will not be able to dissolve any more drink crystals because it is saturated with them. A solution is **saturated** with a solute when no more of the solute can be dissolved in it. A solution is **unsaturated** with a solute when more of the solute can be dissolved in it.

Figure 3
How much sugar do you think can be dissolved in this lemonade?

The ability of a substance to dissolve in a ~~solute~~ solvent is called **solubility.** You can measure the exact amount of solute that is required to form a saturated solution in a certain solvent at a certain temperature. Temperature is important because you can generally dissolve more solute in warm water than in cold water.

Solubility is different for each combination of solute and solvent. The amounts of different solutes that are needed to saturate a certain volume of solvent varies enormously. For example, more sugar than salt is needed to saturate 100 mL of water at room temperature (20°C) (**Table 1**).

Table 1 Solubilities of Common Substances in Water

Solute	Temperature (°C)		
	0	**20**	**50**
baking soda	6.9 g/100 mL	9.6 g/100 mL	14.5 g/100 mL
table salt	35.7 g/100 mL	36.0 g/100 mL	36.7 g/100 mL
sugar	179 g/100 mL	204 g/100 mL	260 g/100 mL

TRY THIS: *DISSOLVE SOLUTES*

Skill Focus: predicting, observing

1. Make a saturated solution of water and salt by stirring small amounts of salt into about 100 mL of water until no more salt will dissolve.

2. Now that the water is saturated with salt, do you think you will be able to dissolve anything else in the water? Make a prediction.

3. Test your prediction by trying to dissolve sugar in your saturated salt solution.

Supersaturation

A very few solid solutes can be used to create a solution that is more than saturated. A solution that contains more of the solute than can be found in a saturated solution is called a **supersaturated** solution.

You can make a supersaturated solution by starting with a saturated solution at high temperature and then allowing the solution to cool slowly. Normally, as a solution cools, the solute particles lose energy. Some of the solute particles draw together and form the crystal pattern of the solid. In a supersaturated solution, the solute particles are not able to get into a crystal pattern. As a result, the solution remains liquid even when it is at a temperature at which it would normally be a solid.

If the supersaturated solution is not disturbed, all the solute may remain dissolved. If you strike the container lightly with a stirring rod or a spoon, however, the resulting vibrations may cause some of the solute particles to move into a crystal pattern. Immediately, the rest of the extra solute will fall out of solution and join the crystal. You can produce a similar effect by adding a seed crystal of the solute for the excess solute particles to build on (**Figure 3**).

LEARNING TIP ◁

Pause and think. Ask yourself, "What did I just read? What did it mean?" Try to reword the information on supersaturation in your own words.

Figure 3
Adding a seed crystal causes the rapid formation of crystals in a supersaturated solution of sodium acetate.

LEARNING TIP ◁

Locate the information needed to answer these questions by scanning the text for key words.

⫸ CHECK YOUR UNDERSTANDING ⊗

1. Identify the solute and the solvent in the photo to the right.

2. List two liquid solutions that do not contain water.

3. Suppose that you add one teaspoon of sugar to your cup of tea. A friend adds four teaspoons of sugar to his cup of tea. Whose tea is a more concentrated sugar solution?

4. Is the solubility of all solutes the same? Explain.

5. Describe how you can tell the difference between an unsaturated solution and a saturated solution.

○ SKILLS MENU

○ Questioning	● Observing
● Predicting	● Measuring
● Hypothesizing	○ Classifying
● Designing Experiments	○ Inferring
● Controlling Variables	● Interpreting Data
○ Creating Models	● Communicating

▷ **LEARNING TIP**

For help in writing hypotheses, controlling variables, or writing up your experiment, see the Skills Handbook sections "Hypothesizing," "Controlling Variables," and "Writing a Lab Report."

sugar cubes

beaker

water

Factors That Affect the Rate of Dissolving

Imagine that you are trying to make a cold drink, such as lemonade, in a hurry. All that you have to sweeten your drink are sugar cubes, and they seem to be taking forever to dissolve. When you add sugar to a drink, several factors affect how quickly the sugar dissolves. Based on your experiences, you probably have some ideas about what these factors are. But have you ever tested your ideas?

Question

What factors affect how quickly a solute dissolves in a solvent?

Hypothesis

You will be testing at least three variables in this experiment. Write a hypothesis for each variable that you plan to test. Use the form "If … then the sugar will dissolve more quickly."

Materials

- sugar cubes
- beakers
- water

List any other materials that you will need to perform this experiment.

▶ Procedure

- This is a controlled experiment to investigate factors that affect the rate of dissolving. Design a procedure for the experiment. For each part of your procedure, determine
  - what variable(s) will change
  - what variable(s) will stay the same
- Submit your procedure, including any safety precautions, to your teacher for approval. Also submit a diagram, at least half a page in size, showing how you will set up your experiment.

Data and Observations

Create a table to record your observations. Record your observations as you carry out your experiment.

Analysis

1. Compare your results with your classmates' results. Were your results similar?

2. Use the particle model of matter to explain why each factor affected the rate of dissolving.

Conclusion

Were your hypotheses correct? Did your observations support, partly support, or not support your hypotheses? Write a conclusion for your experiment.

LEARNING TIP ◁

For help in writing a conclusion, see the example and explanation in the "Writing a Lab Report" section of the Skills Handbook.

Applications

1. A soup recipe calls for bouillon to be added. You find both bouillon powder and bouillon cubes in your kitchen cupboard. Which form of bouillon will speed up the soup-making process? Explain your answer.

2. Most brands of soda pop are solutions of water, dissolved sugar, and dissolved carbon dioxide gas. When you remove the cap from a cold bottle of pop, you hear a faint whoosh as the gas escapes. When you remove the cap from a warm bottle, however, the whoosh is much louder. What effect do temperature and pressure have on the rate that carbon dioxide gas comes out of a pop bottle?

3. Give three examples of situations in which speeding up or slowing down the rate of dissolving might be important. How do you think this could be done in each situation?

⫸ CHECK YOUR UNDERSTANDING ⊗

1. What were the independent and dependent variables in each part of your procedure?

2. Suggest at least two factors that you think would have no effect on the rate of dissolving. Explain why you think they would have no effect.

6.8 Measuring the Acidity of Solutions

Have you ever wondered what makes lemon juice sour (**Figure 1**)? Lemon juice is a solution that contains dissolved compounds. Scientists classify some compounds by the properties of the solutions they form.

Figure 1
Lemon juice has a sour taste.

▷ **LEARNING TIP**

Check your understanding of the properties of acids and bases. Work with a partner and take turns describing the properties.

Acids are compounds that form solutions with the following properties:

- have a sour taste
- react with (corrode) metals
- can cause serious burns on skin

Many acidic solutions, such as lemon juice and vinegar, are harmless. They can be used to give foods a tangy flavour. Other acidic solutions are extremely dangerous. Hydrochloric acid, for example, is used to etch concrete and would make holes in your skin or clothing.

Bases are compounds that form solutions with the following properties:

- have a bitter taste
- feel slippery
- react with (break down) fats and oils
- can cause serious burns on skin

Some basic solutions are harmless. You can drink a solution of baking soda and water to calm an upset stomach. Other basic solutions, such as drain cleaner, should be used with extreme care. They should never be allowed even to touch your skin.

You often use the properties of acidic and basic solutions in your daily life. Some common acids and bases are shown in **Figure 2**.

LEARNING TIP ◁

Make connections between what you are learning about the properties of acids and bases in this section and what you already knew about the products in **Figure 2**. Ask yourself, "Did I already know these products had these properties?"

Figure 2
Common acids (left) and bases (right)

Identifying Acids and Bases

Because many acids and bases are not safe to taste, scientists use other properties to identify them. One property that is safe to use is their effect on a dye called litmus [LIHT-muhs]. Acidic solutions turn blue litmus paper red. Basic solutions turn red litmus paper blue. Litmus is called an indicator because it indicates whether a solution is acidic or basic (**Figure 3**).

Figure 3
Blue litmus paper turns red in an acidic solution (left). Red litmus paper turns blue in a basic solution (right).

Scientists measure acidity on the **pH** scale—a scale of numbers running from 0 to 14 (**Figure 4**). If a compound is neither an acid nor a base, it is **neutral** and has a pH of 7.0. Pure water, for example, is neutral. **Acidic** solutions have pH values that are below 7. The more acidic a solution, the lower its pH value is. A solution with a pH between 0 and 3 is very acidic. **Basic** solutions have pH values that are above 7. The more basic a solution, the higher its pH value is. Very basic solutions, such as drain cleaners, have pH values that are close to 14.

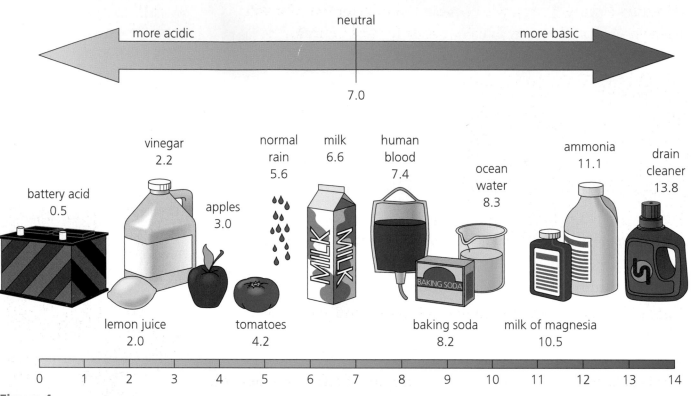

Figure 4

The pH values of some common substances

> **▶ CHECK YOUR UNDERSTANDING**   ⊗
>
> **1.** Make a Venn diagram to compare acids and bases. Include at least two examples for each.
>
> **2.** When might you need to know whether a solution is acidic or basic?
>
> **3.** Dishwasher detergent, oven cleaner, and drain cleaner are all basic solutions with high pH values. What property of basic solutions makes these products useful?

Measuring the pH of Household Products

Acids and bases can be distinguished from one another by the colours they turn certain indicators. The pH paper you will use in this investigation is a universal indicator. It contains several indicators, which turn different colours in solutions with different pH values (**Figure 1**). You will also use an indicator that you will make yourself and develop a colour scale for this indicator.

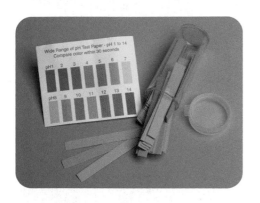

Figure 1
pH paper turns different colours.

Question

What is the pH of some common products around your home?

Materials

- red cabbage
- 400-mL beaker
- hot plate
- apron
- safety goggles
- several clean beakers or glass jars
- several household products that are solutions, such as lemon juice, apple juice, baking-soda solution, vinegar, milk, coffee, tea, and shampoo
- samples of different types of water, such as tap water, rainwater, swimming-pool water, hot-tub water, and water from a lake or stream
- wide-range pH indicator paper
- medicine dropper

 Acids and bases are harmful to eyes and skin. Always wear safety goggles and an apron.

Procedure

1 Slice the red cabbage, and put it in the 400-mL beaker. Add water to cover the cabbage. Boil the cabbage on a hot plate for about half an hour while you continue with the other steps. After half an hour, turn off the heat and let the cabbage cool.

✋ **Be careful when using the hot plate.**

2 In your notebook, draw a data table like the one below.

Data Table for Investigation 6.9		
Substance tested	Prediction: acid, base, or neutral?	pH
lemon juice		
baking-soda solution		

3 Before you test the solutions, predict whether each solution will be acidic, basic, or neutral. Record your predictions in the second column of your table.

4 Put on your apron and safety goggles.

5 Put about 20 mL of a different household product or different type of water in each beaker.

✋ **There is the risk that acids and bases may irritate eyes and skin. If you get any acid or base in your eyes or on your skin, immediately rinse the area with water for 15 to 20 min, and tell your teacher.**

6 Test the pH of each solution by dipping a strip of pH paper into the beaker. Compare the colour that results with the colour scale on the dispenser. If the colour that results is in between two colours on the scale, estimate the pH. Record the pH for each solution in your data table.

7 Add 5 drops of red cabbage juice to each solution you tested in step 6. Record the colours. Use the information from the data table you completed in step 6 to develop a red cabbage indicator scale. Set up your indicator scale in a table like the one below.

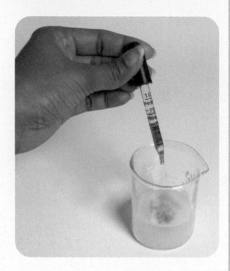

Red Cabbage Indicator Scale	
pH	Colour of red cabbage indicator
1	
2	
3	
4	
5	
6	

8 After you have completed your tests, wash your hands.

Analyze and Evaluate

1. Which solutions were acidic? Which were basic? Which were neutral?

2. Which solution was the most acidic? Which was the most basic?

3. Why do different samples of water have different pH values?

Apply and Extend

4. Samples of water are taken from two swimming pools. One sample has a pH of 4. The other sample has a pH of 5. Which is more acidic?

5. Some people like to squeeze a few drops of lemon juice into their tea (**Figure 2**). When they add the lemon juice, the tea changes colour. Use what you know about acids and bases to explain why this happens. What other substances might cause the colour of tea to change?

Figure 2
What happens to tea when lemon juice is squeezed into it?

LEARNING TIP ◁

For a review of what is involved in making predictions in science, see "Predicting" in the Skills Handbook.

▌▶ *CHECK YOUR UNDERSTANDING* ⊗

1. Did you have any surprises in your predictions about acids and bases? What have you learned that would help you make more accurate predictions if you were given a new set of samples?

LEARNING TIP

For more information on the steps in exploring an issue, see "Exploring an Issue" in the Skills Handbook.

Should Salt Be Used on a Walkway?

Different substances have different properties because every pure substance is made up of different kinds of particles. Each pure substance has its own melting point and boiling point. Adding a solute to a substance changes the melting point or boiling point of the substance.

In winter, snow and ice can make driving and walking dangerous (**Figure 1**). To melt the ice and reduce the danger, salt (sodium chloride) is often spread on roads and walkways. Why is this done? What does salt do to the snow and ice to help make the roads and walkways safer?

Figure 1
How could spreading salt on this road make it safer for drivers?

Salting roads and walkways makes driving and walking safer, but it also causes some serious problems. What are these problems? Are there alternatives to using salt on roads and walkways?

The Issue

It is the middle of winter and the walkway in front of your school has completely frozen over. You are worried that someone might slip on the ice. What should you do? Should you spread salt on the ice? How does this help? What are the benefits of spreading salt on the ice? What are the drawbacks? Are there alternatives to salt? What are the benefits and drawbacks of these alternatives?

Background to the Issue

Gather Information

Work in pairs to learn more about spreading salt on walkways. Where can you find more information? If you are doing an Internet search, what key words can you use?

www.science.nelson.com GO

LEARNING TIP

For more information about how to do this research, see "Researching" in the Skills Handbook.

Identify Solutions

You may wish to use the following questions to help you identify solutions:

- What does salt do to ice and snow? (*Hint*: Think in terms of chemistry—melting point, solutions, and so on.)
- What are the drawbacks of using salt on walkways?
- What are some alternatives to using salt?
- How much does each alternative cost?
- What are the benefits and drawbacks of each alternative?
- What are the environmental impacts of salt and each of the alternatives?

Make a Decision

What will you use on the walkway? What criteria did you use to decide?

Communicate Your Decision

Write a position paper about why salt should or should not be used on the walkway. If you decide that salt should not be used, explain why. Then discuss alternatives to salt. Explain which alternative(s) you would use, and why.

CHECK YOUR UNDERSTANDING

1. How did you come up with your position? Are there things you could have done differently? Explain.
2. Why should you always be prepared to consider alternatives?
3. Why should you always be prepared to defend your position on an issue?

Key Idea: **All matter can be classified as pure substances or mixtures.**

Vocabulary

pure substance p. 137

mixture p. 138

pure substance

pure substance

mixture

Key Idea: **Pure substances can be classified as elements or compounds.**

Vocabulary

elements p. 140

compounds p. 141

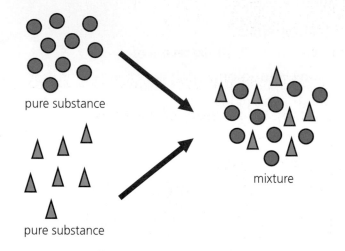

Element

Compound

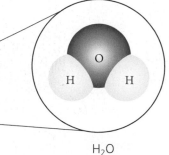

H_2O

Key Idea: Mixtures can be classified as mechanical mixtures, suspensions, or solutions.

Mechanical Mixture

Suspension

Solution

Key Idea: Mixtures can be separated by a variety of methods:

Evaporating

Using density

Filtering

Dissolving

Picking apart

Using magnetism

Key Idea: Solutions can be measured by concentration, solubility, and acidity.

Concentration

Solubility

Acidity

Review Key Ideas and Vocabulary

When answering the questions, remember to use vocabulary from the chapter.

1. What is the difference between a pure substance and a mixture? Give two examples of each.

2. What is the difference between an element and a compound? Give two examples of each.

3. Identify each of the following as a mechanical mixture, a suspension, or a solution. Explain your reasoning.
   a) granola
   b) orange juice
   c) tap water
   d) a toonie
   e) farm-fresh milk
   f) homogenized milk
   g) concrete
   h) clear apple juice
   i) hand lotion
   j) cereal and milk

4. Describe how you would separate the parts of each mixture. What property of matter makes the separation method work? Explain.
   a) sand and salt
   b) dust in a fluffy blanket
   c) sawdust and sand
   d) pebbles and sand
   e) flour and water

5. Read the following statements. Rewrite any statements that are incorrect so that they are correct.
   a) If a solution is saturated at 20°C, it will also be saturated at 25°C.

   b) When some solvent evaporates, a solution becomes more saturated.
   c) When a saturated solution is cooled, some crystals begin to appear in the solution. The solution is now unsaturated.
   d) A solvent is a liquid that dissolves sugar.
   e) A solute is always a solid.
   f) Oil is insoluble.

6. Five solutions have pH values of 3, 5, 7, 9, and 11. State which solution(s) is (are)
   a) acidic
   b) most acidic
   c) neutral
   d) basic
   e) sour tasting
   f) most helpful in breaking down oils and fats

Use What You've Learned

7. Mixtures and compounds both contain two or more elements. How do mixtures differ from compounds?

8. Screens and filters work in the same way. The screens on your windows and doors separate insects (such as flies and mosquitoes) from the air. Make a table like the one below. In the first column, list different types of filters and screens that can be found in your home and school, and in a car. In the second and third columns, identify what is let through and what is held back.

#8. Screen or filter	What is let through	What is held back

9. Imagine that you have spilled a whole bottle of expensive perfumed oil into a bath. What steps could you take to recover as much of the oil as possible?

10. How could you use a flashlight to distinguish between a solution and a suspension?

11. a) Make a table that has three columns. In the first column, list 10 liquids in your home. Determine the substances in each liquid by reading the label on the container.
    b) In the second column of your table, identify the liquids that meet the definition of a solution.
    c) In the third column, list the solvent and the solute(s) in each solution.

#11.	Liquids in home	Is it a solution?	Solvent and solute

12. The label on a large bottle of liquid laundry detergent states that the bottle contains enough detergent to wash 100 loads of laundry. The label on a different brand, in a smaller bottle, also states that the bottle contains enough detergent to wash 100 loads of laundry. Both claims are true. Explain how this is possible.

13. Using what you know about solutions, predict three ways that you could shorten the time a sugar cube takes to dissolve in a drink. Explain your predictions.

14. Oil spills that occur near shorelines are often cleaned up with the help of powerful detergents. What properties of oil and a detergent solution make this work?

Think Critically

15. Is ocean water a saturated or unsaturated saltwater solution? Explain your answer.

16. "A chemical that can dissolve in water is more dangerous than a chemical that cannot." Do you agree with this statement? Explain.

Reflect on Your Learning

17. Print the following terms on sticky notes: suspension, element, matter, mechanical mixture, compound, pure substance, solution, mixture. Arrange the sticky notes on a piece of paper. Draw lines between them to show how scientists classify matter. Check your work by comparing it with the graphic organizer you drew in section 6.3.

Making Connections

Comparing and Contrasting Properties of Substances

Looking Back

In this unit, you have learned how properties of matter are used to classify matter and to distinguish between different substances. You have learned how to use your senses to observe some properties and how to perform simple tests and measurements to investigate other properties.

In this activity, you will use what you have learned to compare and contrast five substances based on their properties.

Demonstrate Your Learning

Part 1: Lab Work

In this part of the activity, you will work with a partner to explore the properties of five different substances.

Materials

- safety goggles
- apron
- 5 100-mL beakers
- 50 mL of each of the following substances: corn starch, baking soda, washing soda, Epsom salts, and borax
- 7 250-mL beakers
- tap water
- set of measuring spoons
- paper towel
- 5 strips of pH paper
- vinegar
- pan
- scissors
- white glue
- food colouring

Procedure

1. Read through the procedure to find out what you will be doing. Work with your partner to design a data table for recording your observations and results. You and your partner should each make your own copy of the data table.

 Wear safety goggles and an apron throughout this activity.

2. Put on your safety goggles and apron.

3. Label five 100-mL beakers "cornstarch," "baking soda," "washing soda," "Epsom salts," and "borax." Put 50 mL of each substance in the correct beaker.

4. Observe each substance. Record the observable properties of each substance in your data table.

 Do not taste any of these substances. Some of them are not safe to taste.

5. Fill five 250-mL beakers with 200 mL of tap water. Label the beakers "cornstarch and water," "baking soda and water," and so on. Check whether each substance is soluble in water by trying to dissolve 5 mL (1 level teaspoon) of the substance in the correct beaker. Wipe the spoon with a paper towel between substances. Record your results.

6. Test and record the pH of each solution you made in step 4.

7. Set aside the beaker of borax and water for Part 2.

8. Put 10 mL (2 level teaspoons) of the washing soda and water solution in each of the three remaining solutions: cornstarch and water, baking soda and water, and Epsom salts and water. Record your results.

9. Set aside any beaker in which there is a change so that you can observe it again the next day.

10. Empty the remaining solutions into the sink. Remove the labels, and wash the empty 250-mL beakers.

11. The next day, observe the beakers you set aside. Record your observations.

12. Empty the beakers, except the borax/water beaker you set aside in step 7, in the sink and wash them.

Part 2: Just for Fun

You will not be assessed on this part of the activity.

1. *Oobleck*: Use your leftover cornstarch to make oobleck. Add small amounts of water to the cornstarch, stirring gently, until the mixture is the consistency of pancake batter. The mixture should "tear" if you run a finger through it and then "melt" back together. Slowly pour some of the oobleck into a pan. As you pour it, try to cut the stream with scissors. When all the oobleck is in the pan, try hitting it with your hand. Is it solid or liquid? Try putting your hand into it slowly. Is it solid or liquid? Roll some oobleck between your hands to make a ball. Does it hold its shape when you stop? Make a "worm," and pull it apart quickly. What happens? When does oobleck act like a solid? When does it act like a liquid? Clean up the materials you used with warm water.

13. Fill five 250-mL beakers with 100 mL of vinegar. Label the beakers "cornstarch and vinegar," "baking soda and vinegar," and so on. Check whether each substance reacts with an acid (vinegar) by adding 5 mL (1 teaspoon) of the substance to the correct 250-mL beaker. Record your results.

14. Empty all the vinegar solutions into the sink. Remove the labels and wash the beakers.

15. Return any unused portions of the solid substances (except the cornstarch) to the containers provided by your teacher. Remove the labels and wash the beakers.

2. *Bouncy slime balls*: Get a 250-mL beaker. Mix together 30 mL (2 level tablespoons) of white glue, 20 mL (4 level teaspoons) of water, and 3 drops of food colouring in the beaker. Add 30 mL (2 level tablespoons) of the borax and water solution you set aside earlier. Stir. Take the resulting substance out of the beaker. Divide it between you and your partner. See who can stretch it the farthest. Can you break it? Will it bounce? In what ways is it similar to oobleck? Clean up the beakers and measuring spoons with warm water.

3. Wash your hands thoroughly.

Part 3: Analysis of Results

In this part of the activity, you will work on your own. Use the information in your data table for Part 1 to answer the following questions.

1. Did you see any chemical changes in Part 1? If so, what combination(s) produced each change? What clues showed that a chemical change had occurred?

2. Do any of the five substances have a single property that distinguishes them from all the other substances? If so, which substances are they and what is the property?

3. Choose any two of the five substances. Make a Venn diagram to show which of their properties are the same and which are different. Staple your data table and Venn diagram together, and give them to your teacher.

▐▶ ASSESSMENT

LAB WORK

Check to make sure that your lab work provides evidence that you are able to

- work cooperatively with a lab partner
- work safely in a lab situation
- follow procedures
- measure with precision
- record observations and measurements
- clean up lab equipment

DATA TABLE

Check to make sure that your data table provides evidence that you are able to

- design a data table
- identify observable properties
- record observations and measurements accurately
- communicate clearly

ANALYSIS OF RESULTS

Check to make sure that your analysis of results provides evidence that you are able to

- identify properties that are common to two or more substances
- identify properties that are unique to a substance
- identify chemical changes
- use appropriate science terminology
- communicate clearly

EARTH'S CRUST

Preview

Does the land around your community look like the land in the photograph? If not, what land features are found on Earth's surface around your community? Are there mountains, hills, or cliffs? Are there lakes, rivers, or streams? Are you near the ocean? Have you ever noticed any changes in the land features? Do local Aboriginal peoples tell any stories about these land features?

Earth has not always looked the way it does today. Many of the land features that you see every day—including rocks, rivers, and hills—have changed during Earth's long history. In fact, they are still changing. Some of these changes happen rapidly, while others occur over long periods of time.

In this unit, you will learn about changes to Earth's surface and what causes these changes. It is not possible to bring large land features into the classroom to study. It is also not possible to stay in grade 7 long enough to study changes to Earth's surface that take millions of years! Instead, you will create and use models to study features of Earth's surface that are too large to bring into your classroom. As well, you will create and use models to demonstrate, in a few minutes, changes that normally take millions of years.

TRY THIS: OBSERVE LOCAL LAND FEATURES

Skills Focus: observing, inferring

Go outside and observe your community. On a large piece of paper, sketch all the land features in your local area. Remember to include mountains, hills, cliffs, valleys, rivers and streams, and other bodies of water. Beside each feature, write a few brief notes to summarize any stories or legends you have heard about this feature, how you think this feature might have been formed, and how it might be slowly changing. Put the paper in your notebook so that you can add to it during the unit.

◀ This photo was taken at Racing River in northern British Columbia.

CHAPTER 7

Old rocks can be recycled into new rocks.

KEY IDEAS

▶ Minerals are the building blocks of rocks and can be identified by their physical properties.

▶ Earth's crust is made up of three families of rocks: igneous, sedimentary, and metamorphic.

▶ Fossils provide evidence of changes in life over time.

▶ Rock materials are broken into smaller pieces by mechanical, chemical, and biological weathering.

▶ Weathered materials are moved from one place to another by gravity, wind, water, and ice in a process called erosion.

▶ Rocks and weathered rock materials can be transformed into new rocks.

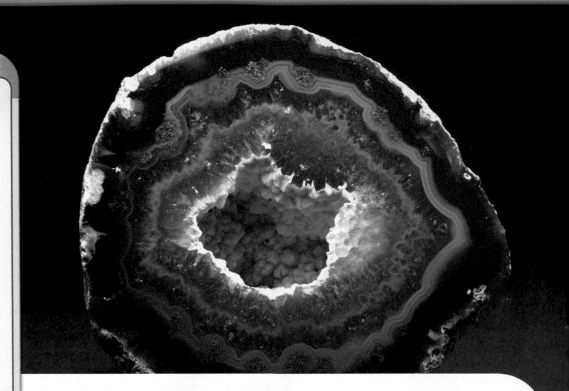

Rocks are all around you. You have even eaten rocks. Those tiny white grains in the saltshaker may be ground-up particles of rock salt, mined from the ground. Materials in everyday products—from glassware, toothpaste, baby powder, and pencils to buildings, automobiles, and computers—come from rocks. Rocks can be used in so many different ways because the materials that make up rocks have different properties.

Some ordinary rocks you see may be very beautiful inside, like the geode above. How did such a beautiful rock form? Why are there layers and different colours? In this chapter, you will investigate what rocks are made of and learn how to classify different types of rocks. As well, you will look for clues in rocks that tell you how they were formed. You will look at evidence that shows how rocks have changed in the past and are still changing today. If you think of rocks as solid and unchanging, you may be in for a surprise! Earth is really a huge rock recycler.

Minerals: Building Blocks of Rocks

If you look carefully at rocks, you will soon see that they are not all the same. Some are white, and some black. Others are brightly coloured or have several different colours. One rock may be soft and dull, whereas another rock may be hard and shiny. An important step in learning how to understand rocks is finding ways to classify them into groups based on their properties. Properties are observable facts about a material, such as colour.

TRY THIS: OBSERVE ROCKS

Skills Focus: observing

Find two or three small, interesting-looking rocks in your schoolyard or at home. Each rock should have some features that make it different from the other(s). Examine and record the properties of your rocks.

1. Make a chart to organize the properties of rocks that you observe.

Properties	Rock 1	Rock 2	Rock 3
colour			

2. What colour(s) are the rocks?
3. Do they look the same throughout, or do they have different types of materials mixed together?
4. Do they feel heavy or light in comparison to their size?
5. Do they have pieces that sparkle or reflect light?
6. Which of your rocks is the hardest? How can you tell?
7. Do your rocks look like most other local rocks? If not, why do you think they are different? How do you think they got to where you found them?

All rocks are made of minerals. **Minerals** are pure, naturally occurring substances that are found in Earth's crust. Do you know someone who wears a diamond ring? Diamonds come from rocks. The graphite in your pencil is a mineral (**Figure 1**). You can think of minerals as the "building blocks" of rocks.

Scientists who study, identify, and classify rocks are called geologists [gee-OL-o-gists]. On the next few pages, you will learn about some of the properties that geologists use to identify the minerals that make up rocks.

Figure 1
Graphite is used to make pencils.

 **LEARNING TIP**

Preview the next four pages. Each heading is a property. Under each heading, the property is explained and you are given examples in both words and photographs. Make notes using this structure.

Colour

Colour is easy to determine and can be an important clue to a mineral's identity (**Figure 2**). By itself, however, colour is not a reliable way to identify minerals. Different minerals may be the same colour. For example, both gold and pyrite (fool's gold) are yellow. Some minerals occur in many different colours. For example, quartz is often white, but it can also be violet, gray, black, or colourless (**Figure 3**).

Figure 2
Jade is usually a shade of green

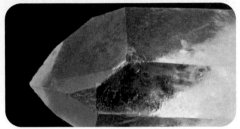

Figure 3
Quartz is sometimes colourless.

Streak

Streak describes the colour of the powdery mark that some minerals make when they are scratched against a hard surface. To see the streak clearly, geologists scratch a mineral on a streak plate. A streak plate is an unpolished piece of porcelain tile. The colour of the streak may be the same as the colour of the mineral, or it may be different. The colour of the streak is more reliable than the colour of the mineral. For this reason, it is very useful for identifying some minerals. For example, hematite can be shiny silver or reddish, but it always has a reddish streak (**Figure 4**). Pyrite (fool's gold) and gold are both yellow, but gold makes a yellow streak and pyrite makes a dark streak (**Figure 5**).

Figure 4
Different colours of hematite make the same colour streak.

Figure 5
During the gold rush, prospectors used streak to test if they had found real gold.

Lustre

Lustre [LUST-er] is the degree of shininess. Some minerals, such as gold, have a metallic lustre (**Figure 6**). Others, such as obsidian, look glassy (**Figure 7**). Still others, such as asbestos, have a dull appearance (**Figure 8**).

Figure 6
Gold has a metallic lustre.

Figure 7
Obsidian has a glassy lustre.

Figure 8
Asbestos has a low lustre.

Hardness

The hardness of a mineral can be determined by scratching one mineral against another. A mineral can make a scratch on any mineral that is softer than it is, but it cannot make a scratch on a mineral that is harder than it is. Geologists use a set of 10 standard minerals, ranging from very soft to very hard, to compare hardness. This is called the Mohs hardness scale (**Table 1**) after Friedrich Mohs (1773–1839), the German scientist who developed it. If you cannot obtain a set of Mohs hardness scale minerals, you can make your own using everyday materials.

Table 1 Scale for Comparing the Hardness of Minerals

Mohs hardness scale		Hardness scale of materials you can easily find	
1 talc	*SOFTEST*	**1**	soft pencil point
2 gypsum		**2–3**	fingernail
3 calcite			
4 fluorite		**3–4**	copper penny
5 apatite		**5–6**	nail or glass bottle
6 feldspar			
7 quartz		**6–7**	steel file
8 topaz		**7–8**	sandpaper
9 corundum		**9**	emery paper
10 diamond	*HARDEST*		

Crystal Structure

All minerals are crystals (**Figures 9** and **10**). Crystals have regular shapes because they are made up of tiny particles that are connected in a repeating pattern. The size of the crystals tells geologists how quickly a mineral cooled from a liquid to a solid. Large crystals indicate that the mineral cooled slowly. Small crystals indicate that the mineral cooled rapidly. Most crystals are too small to be seen without magnification.

Figure 9
Crystals of wulfenite

Figure 10
Crystals of pyrite

Cleavage

Some minerals break, or fracture, into pieces with rough, uneven surfaces. Quartz breaks in this way. Other minerals usually split or crack along parallel or flat surfaces. This property is called cleavage. You can test a mineral by breaking it with a hammer or splitting off sheets with a dinner knife. For example, mica (**Figure 11**) always splits into thin sheets. Other minerals, such as halite (**Figure 12**), always split into cubes.

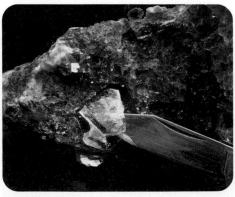

Figure 11
Mica

Figure 12
Halite (also called table salt)

Magnetism

Magnetism is the ability of a mineral to attract a magnet. Only minerals that contain iron are magnetic, so most minerals are not magnetic. You can use a magnet to find out if a mineral is magnetic (**Figure 13**).

Figure 13
Magnetite will attract or repel a magnet.

Reaction with Certain Chemicals

Some minerals can be identified by their reaction with certain chemicals. For example, calcite, limestone, and marble react with acidic solutions, such as vinegar (**Figure 14**). The acidic vinegar reacts with the carbonate materials in these minerals, creating a fizzing or bubbling on the surface. The gas that fizzes or bubbles up is carbon dioxide.

Figure 14
Limestone fizzes as it reacts with vinegar.

⫸ CHECK YOUR UNDERSTANDING

1. List the eight properties that are used to classify minerals.

2. What is one advantage of using colour to identify a mineral? What is one disadvantage of using colour to identify a mineral?

3. Why do geologists use both the colour of a mineral and the colour of its streak to identify the mineral?

4. Why do you think geologists use drill bits covered with small diamonds to drill into Earth's crust?

LEARNING TIP ◁

Do not guess. Look back through the section to find the answers. Even if you remember the answer, it is good to go back and check it.

7.1 Minerals: Building Blocks of Rocks

Identifying Minerals

In this investigation, you will identify several different minerals by examining some of the properties you have just learned about (**Figure 1**). Geologists use these properties, as well as others, to identify minerals.

SKILLS MENU

- ○ Questioning
- ● Observing
- ○ Predicting
- ○ Measuring
- ○ Hypothesizing
- ● Classifying
- ○ Designing Experiments
- ○ Inferring
- ○ Controlling Variables
- ● Interpreting Data
- ○ Creating Models
- ● Communicating

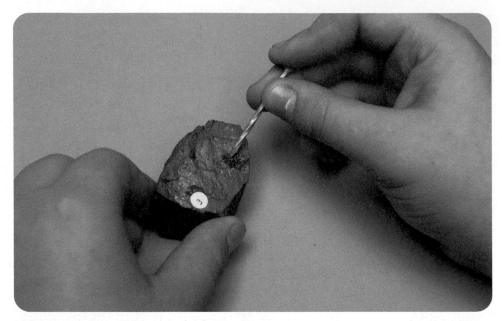

Figure 1
Hardness tests can be used to help identify unknown minerals.

Question

Can you identify unknown minerals by their properties?

Materials

- safety goggles
- apron
- set of numbered mineral samples
- Mohs hardness scale set of minerals (or substitutes)
- hand lens
- streak plate (unglazed tile)
- magnet
- dropper
- vinegar
- hammer

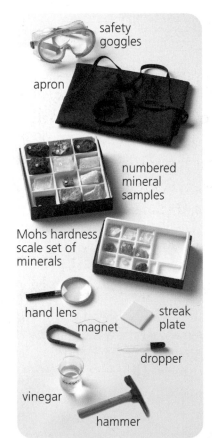

safety goggles

apron

numbered mineral samples

Mohs hardness scale set of minerals

hand lens

magnet

streak plate

dropper

vinegar

hammer

▶ Procedure

1 Copy the table below into your notebook.

	Mineral number	Colour	Streak	Lustre	Hardness	Magnetism	Reaction with vinegar	Cleavage	Name
	1	grey–black	reddish brown	metallic					

Data Table for Investigation 7.2

2 Select a mineral from the set your teacher provides. Record the number of the mineral in the first column of your table.

3 *Colour:* What colour is your mineral? Record the colour of your mineral in your table.

4 *Streak:* Rub your mineral across the streak plate. Brush off the extra powder with your fingers. Record the colour of the streak, if any.

5 *Lustre:* Is the lustre of your mineral metallic (like polished metal) or non-metallic? Is it brilliant, glassy, pearly, silky, waxy, or dull? Try to find the best words to describe the lustre of your mineral, and record your observations.

6 *Hardness:* Scratch your mineral with Mohs mineral #5 (or a nail). If this does not leave a scratch or groove in your mineral, continue along the scale toward #10. If mineral #5 does leave a scratch in your mineral, move along the scale toward #1. Rank the hardness of your mineral. (It will be between two numbers unless your mineral is identical to one of the minerals in the Mohs hardness scale.) Record your results.

7 *Magnetism:* Use a magnet to determine if your mineral is magnetic. Record your result.

8 *Reaction with vinegar:* Use the dropper to put a few drops of vinegar on your mineral. Does it fizz? Record your results.

 Wear safety goggles during step 9.

9 *Cleavage:* Your teacher will use a hammer to break your mineral. Does it break along flat surfaces, or does it break into pieces with rough, uneven surfaces? Record your observations.

10 Repeat steps 2 to 9 for the rest of the minerals in your set.

11 Use **Table 2** to help you identify your minerals. If you can identify a mineral, write its name in the last column of your data table. If you cannot identify a mineral, write "do not know."

Table 2 Characteristics of Some Common Minerals

Mineral	Colour	Streak	Lustre	Hardness	Magnetism	Reaction with acid	Cleavage	Other
graphite	black	black	metallic	1–2				slippery feel
galena	grey	grey	metallic	2–3				square corners
halite	white	colourless	glassy	2–3			three cleavage planes	square corners
biotite	black	white/ pale grey	glassy/ brilliant	2–3			splits into leaves; one cleavage plane	
calcite	white	colourless	glassy	3		fizzes		
pyrrhotite	yellow-brown	black	metallic	4	magnetic			
serpentine	shades of green	colourless	silky/waxy	2–5				
magnetite	black	black	metallic	6	magnetic			
hematite	red/black	red	metallic/ dull	5–7				
feldspar	white/ pink/ greenish	colourless	pearly	6			two cleavage planes	
pyrite	yellow	brown/ black	metallic	6–7				
quartz	colourless/ white/rose	colourless	glassy	7			two reflective surfaces	will scratch glass

Analyze and Evaluate

1. Did lustre help you identify any of the minerals in your set? If so, which one(s)?

2. Do you think that streak is more useful for identifying minerals than colour or lustre? Explain your answer using an example.

3. Do you think that magnetism is a useful property for identifying minerals? Suggest a reason for your answer.

Apply and Extend

4. Which properties could you use to identify minerals if you were out for a walk and had no equipment?

⫸ CHECK YOUR UNDERSTANDING

1. Why is it important to wear safety goggles when investigating the cleavage of minerals?

2. Why do you think you were asked to record your observations and results in a table?

Families of Rocks

There are many different minerals, but they are usually found mixed together in rocks. For example, granite contains mica, quartz, and feldspar.

Geologists classify rocks into three families based on how they are formed. These are igneous, sedimentary, and metamorphic rock.

Igneous Rock

Hot molten rock under Earth's surface is called **magma.** Rock that forms from the hardening of liquid magma is called **igneous rock** [IG-nee-us]. Most of Earth's surface is composed of igneous rock, and igneous rock is still being formed today.

If the magma cools underground, the rock that is formed is called **intrusive** igneous rock. This type of igneous rock is seen on Earth's surface only after years of erosion have worn away the layers of rock over it. Stawamus Chief near Squamish, British Columbia is one of the world's largest chunks of granite, a common intrusive igneous rock (**Figure 1**).

If the magma is forced out onto Earth's surface, it is called **lava.** Igneous rock that is formed on Earth's surface when the lava cools is called **extrusive** igneous rock. Basalt is extrusive igneous rock that is common in British Columbia.

Figure 1
The Stawamus Chief is popular with climbers. Many Aboriginal groups have special stories and legends about unique features like the Stawamus Chief.

The formation of both intrusive and extrusive igneous rock is shown in **Figure 2**.

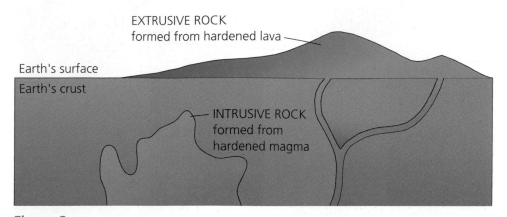

EXTRUSIVE ROCK
formed from hardened lava

Earth's surface

Earth's crust

INTRUSIVE ROCK
formed from
hardened magma

Figure 2
Two types of igneous rock

LEARNING TIP ◁

Are you able to explain the difference between intrusive and extrusive igneous rocks in your own words? If not, ask yourself, "What do I need to figure out? What don't I understand?" Then re-read the explanation, and re-examine **Figure 2**.

The rate at which the molten material cools determines the size of the crystals in the rock. Granite (**Figure 3**) is intrusive rock that is formed when magma cools very slowly within Earth. Granite has a mottled appearance and it contains crystals that can be seen with the unaided eye. Obsidian (**Figure 4**) is an extrusive igneous rock that is formed when lava cools very quickly, forming very tiny crystals that cannot be seen without magnification.

Figure 3
Granite

Figure 4
Obsidian

Some igneous rock even floats. Off certain South Pacific islands, small pieces of pumice can be found floating in the ocean (**Figure 5**). Small pockets of volcanic gases, which are trapped as frothy lava quickly cools, allow this rock to float.

Figure 5
Pumice is a type of igneous rock that floats.

Sedimentary Rock

When bare rock is exposed at Earth's surface, it may be broken into smaller pieces, or particles, in many different ways. These small rock particles are moved from one place to another. Rain and melted snow wash the rock particles into streams and rivers, which then carry the rock particles for many kilometres. The rock particles, along with clay, mud, sand, gravel, and boulders, are called **sediment.** As the water approaches a lake or ocean, and the current slows, the sediment gradually sinks to the bottom (**Figure 6**).

Figure 6
As this satellite photo shows, sediment carried by the Fraser River pours out into the sea.

▷ **LEARNING TIP**

As you study **Figure 7**, ask yourself, "What is the purpose of this illustration? What am I supposed to notice and remember?"

There, on the lake or ocean floor, the sediment gradually piles up in layers. Over millions of years, the enormous weight of the upper layers of sediment presses down on the lower layers. Under the pressure, the lower layers are compacted (pushed together). Dissolved minerals act as a natural cement that hardens the lower layers into rock. Rock that is formed by the breaking down, depositing, compacting, and cementing of sediment is called **sedimentary rock** (**Figure 7**). Sedimentary rock may also contain animal and plant remains that have been deposited along with the sediment.

Figure 7
Sedimentary rock is formed as layers of sediment are added by a river.

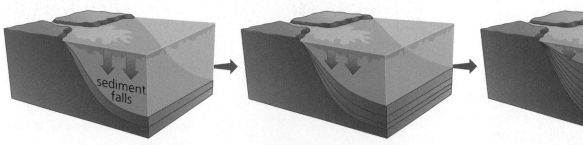

Rock and soil particles and gravel are carried by the river. They sink to the bottom, forming a layer of sediment.

Each new layer puts pressure on the layers below.

Eventually, the lower layers cement into rock.

Figure 8

Shale is a smooth sedimentary rock that is formed from layers of tiny particles of clay or silt.

Sandstone, a rougher rock, is formed from layers of compressed sand.

Conglomerate is made from sediment that contains rounded pebbles and small stones.

As the layers are compressed, they form different kinds of rock, depending on the nature of the particles in the sediment (**Figure 8**).

The appearance and properties of a sedimentary rock can tell you what the original sediment was like. The size of the rock particles that settle to the bottom of a river, lake, or ocean depends on the speed of the water that carried the particles. For example, a narrow, swift-flowing mountain stream on a steep slope can move large rocks. A wide, slow-moving river on flat land can carry only fine clay particles. By studying the layers of sediment in different places today, geologists can understand what the land was like in the past (**Figure 9**).

Although most of Earth's surface is made up of igneous rock, much of the rock you see on the surface is sedimentary rock. New sedimentary rock is being formed all the time as additional layers of sediment are deposited by wind and water.

Figure 9
The layered appearance of these sedimentary rocks is a clue to how they were formed.

TRY THIS: **MAKE YOUR OWN SEDIMENT**

Skills Focus: observing, creating models

Fill a jar with a screw-on lid half full of water. Add some clay, sand, and fine gravel or pebbles. Cap the jar tightly and shake it gently until all the sediment is moving. Put down the jar and observe the sediment settling. What do you notice about the sizes of the particles in each layer?

7.3 Families of Rocks

LEARNING TIP

The word "metamorphic" means "changed in form." It comes from the Greek words *meta*, meaning "after," and *morphe*, meaning "form." The root "metamorph" is used in other areas of science as well. For example, the change in form of a caterpillar into a butterfly is called "metamorphosis."

Metamorphic Rock

When igneous or sedimentary rock becomes buried at a great depth, it is subjected to increased temperature and pressure. As well, magma moving through Earth heats and squeezes the neighbouring rock. As a result, the rock may change. The changed rock is different from the original rock in appearance or in the minerals it contains. Rock formed below Earth's surface, when heat and pressure cause the properties of existing rocks to change, is called **metamorphic rock.**

Some metamorphic rocks have been changed so much that they no longer resemble the original rock, or parent rock. Often, however, geologists can trace the relationship between a metamorphic rock and its parent rock. For example, slate is a metamorphic rock that is formed from the sedimentary rock shale. Gneiss [NICE] is a metamorphic rock that is formed from the igneous rock granite. **Table 1** shows some types of metamorphic rocks and their parent rocks.

Table 1 Metamorphic Rocks

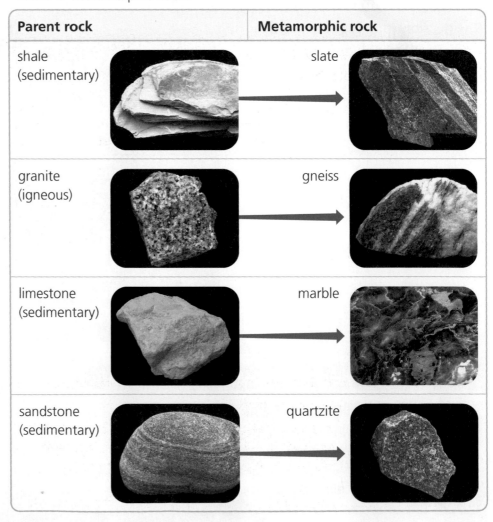

Parent rock	Metamorphic rock
shale (sedimentary)	slate
granite (igneous)	gneiss
limestone (sedimentary)	marble
sandstone (sedimentary)	quartzite

The cycle does not always stop here, however. With more heat and pressure, metamorphic rock can change into other types of metamorphic rock. For example, with additional heat and pressure within Earth, the metamorphic rock slate can change into phyllite, and phyllite can change into schist (**Figure 10**). Schist is one of the strongest rocks in the world. New metamorphic rocks are being formed all the time, deep within Earth.

Slate Phyllite Schist

Figure 10

CHECK YOUR UNDERSTANDING

1. Copy and complete **Table 2**.

Table 2 How Rocks Are Formed

Type of rock	How it is formed	Examples
igneous		
sedimentary		
metamorphic		

2. What is the difference between intrusive and extrusive igneous rock?
3. How are minerals different from rocks?
4. A rock that contains a cavity filled with crystals is called a geode. Look at the photograph at the beginning of this chapter. Why are the crystals in the middle of the geode larger than the crystals toward the outside?

7.4 Fossils

NEWS Flash!! Dinosaur tracks discovered!

Two elementary school boys find evidence of Ankylosaur while tubing down Flatbed Creek.

MARY CHANG

CHARLES HELM

David Helm and Mark Turner discovered fossilized dinosaur footprints near Tumbler Ridge, British Columbia.

In August 2000, two boys went tubing down Flatbed Creek near Tumbler Ridge, British Columbia. As they walked along the banks of the creek, they came across unusual impressions in the rock. What Daniel Helm and Mark Turner had discovered were 20 dinosaur footprints preserved in sedimentary rock that was over 90 million years old. Further investigations by scientists revealed more footprints, as well as fossilized dinosaur bones embedded in the rock only a few metres from the footprints. Across the creek, there were even more footprints.

▷ **LEARNING TIP**

Before you read this section, make a web to show what you already know about fossils.

How Fossils Form

Fossils are rock-like casts, impressions, or actual remains of organisms that were covered by sediment when they died, before they could decompose (**Figure 1**). Only a tiny fraction of organisms are preserved as fossils. This is because most dead organisms decay or are eaten by scavenging animals. Also, soft tissue, such as muscle and body organs, does not fossilize well.

Figure 1
A fossil of a turtle.

An organism that is suddenly buried by falling into mud or quicksand may become a fossil. An organism that is covered quickly by a landslide of sediment or blowing volcanic ash may also become a fossil. The layer of sediment that contains the organism is covered by other layers of sediment and gradually becomes sedimentary rock.

As the wet sediment becomes rock, minerals that are dissolved in the water gradually replace minerals in the body of the buried organism. Minerals in bone, shell, and parts of plants can be replaced this way. Eventually, particle by particle, all the minerals in the organism are replaced by minerals in the water. The final result is a fossil that looks exactly like the original organism but is in a rock-like form (**Figure 2**).

Ammonites were marine animals that looked like octopuses with shells. Although they no longer live on Earth, a close relative, called the chambered nautilus, still does. Vancouver Island and the Gulf Islands are well known for their ammonite fossils, like the one in **Figure 2**. Is there a rocky area near your school where your class could look for fossils?

Figure 2
This ammonite, found in British Columbia, is 170 million years old.

How Fossils Tell Us About Geological Change

Fossils record the history of changes to life on Earth. All the information we have from fossils is called the **fossil record.** The fossil record is important because it shows what types of animals and plants lived on Earth hundreds of millions of years ago. The fossil record also shows how life has changed over time. If you look at exposed layers of sedimentary rock from bottom to top, the fossils are like a series of snapshots of how life has changed on Earth, from the distant past near the bottom to more recent times near the top.

Scientists have used fossils to make a time line of the changes in life on Earth. This time line is called the **geologic time scale** (Figure 3).

Geologists use fossils to compare the ages of rocks. For example, if a certain type of ancient fossil is found in rocks in two different places, then these rocks were probably formed about the same time.

▷ **LEARNING TIP**

Observe how **Figure 3** is organized. You can read it across and down. Each tells a different story.

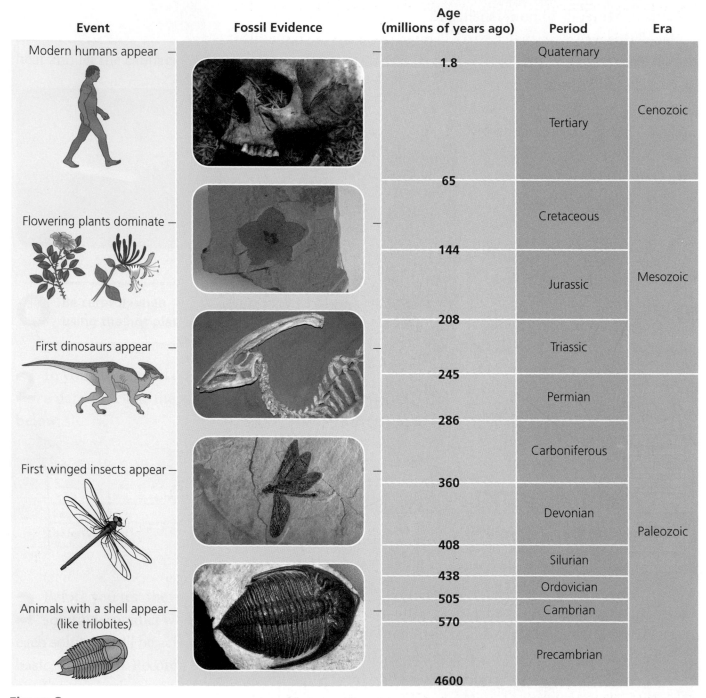

Event	Fossil Evidence	Age (millions of years ago)	Period	Era
Modern humans appear		1.8	Quaternary	Cenozoic
		65	Tertiary	Cenozoic
Flowering plants dominate			Cretaceous	Mesozoic
		144	Jurassic	Mesozoic
First dinosaurs appear		208	Triassic	Mesozoic
		245	Permian	Paleozoic
		286	Carboniferous	Paleozoic
First winged insects appear		360	Devonian	Paleozoic
		408	Silurian	Paleozoic
		438	Ordovician	Paleozoic
Animals with a shell appear (like trilobites)		505	Cambrian	Paleozoic
		570	Precambrian	
		4600		

Figure 3
Geologic time scale

Geologists also use fossils to track how Earth has changed or how parts of Earth have moved over time. The oldest known fossils in Canada are located at the Burgess Shale in Yoho National Park, near Field in eastern British Columbia (**Figure 4**). The fossils here are not dinosaur fossils. They are fossils of soft-bodied marine organisms that existed on Earth 500 million years ago, in a period of geologic time known as the Cambrian era. These fossils are so rare and unusual that the area has been made a World Heritage Site by the United Nations.

The rock that contains these fossils is black shale. Geologists determined that the rock originally formed in the ocean, close to a reef where the marine organisms lived. The reef was at the edge of the ancient North American continent. Occasional underwater mudslides trapped the organisms and, over many millions of years, their bodies were covered by more than 8 km of sediment. Forces inside Earth gradually moved the fossils eastward to their current location, high in the mountains.

Figure 4
The Burgess Shale (left) is one of the most famous fossil finds in the world. This Burgess Shale fossil (right) is so strange that the geologists who first saw it thought they must be hallucinating. They named it *Hallucinogenia*.

▶ CHECK YOUR UNDERSTANDING

1. Describe how fossils form.
2. Why do very few organisms become fossils?
3. Would you expect to find fossils of dinosaur bones in rock that was 250 million years old? Why or why not?
4. Why are fossils found in sedimentary rock but not in metamorphic rock? (Think about how the three families of rock are formed.)
5. Write "Fossils" in the middle of your page. Then make a mind map of all the things that scientists have learned from fossils.

LEARNING TIP ◁

Compare what you learned about fossils in this section to what you already knew about fossils.

7.5 Weathering Breaks Down Rocks

▷ **LEARNING TIP**

Preview the section and read the headings. How many types of weathering will you be learning about in this section?

An old cemetery can be an interesting place to visit. You can see how small, slow changes make a big difference after many years. For example, almost 200 years ago, the gravestone in **Figure 1** was polished and new. Today, the edges are chipped and the surface of the stone has tiny holes. In a few more years, the writing will be worn away. Eventually, the stone will crumble apart and disappear into the soil.

Figure 1
This old gravestone is starting to show wear. Compare it with a new gravestone to see how rock weathers over time.

The process that slowly breaks down natural materials, such as rocks and boulders, into smaller pieces is called **weathering.** Weathering also breaks down human-made structures, such as roads and buildings. Weathering can be caused by physical forces or by chemical reactions.

The term "weathering" indicates that the changes to the rock material are caused by the weather. Weather includes changing temperature, wind, rainfall, and snowfall. Weathering slowly breaks down all rock materials in contact with the air.

There are three kinds of weathering: mechanical, chemical, and biological.

Mechanical Weathering

Weathering that is caused by a physical force is called **mechanical weathering.** Many different physical forces can cause weathering.

Have you ever left a full bottle of water or pop in the freezer? The container either swells or breaks as the water expands from freezing. Rocks often have cracks in them. During colder months, rainwater is caught in the cracks and then freezes. As the water expands, it puts pressure on the walls of the cracks, forcing them to widen. This is called **ice wedging.** Eventually, as ice wedging occurs again and again, the cracks may widen or pieces of rock may break off (**Figure 2**).

LEARNING TIP ◁

Compare what you are learning with what you already know. How do the examples of mechanical weathering fit with what you already know?

Figure 2
Ice wedging has weathered this rock face. Notice the pieces of rock that have fallen off and collected below.

Mechanical weathering may be caused by the wind as well. Sand particles and small rocks carried by the wind have the same effect as sandpaper as they rub the surface of the rock. The sand and small rocks slowly wear down exposed surfaces into small pieces or particles of rock.

Figure 3

These rocks have been rounded as they tumbled in a fast-flowing river.

Similarly, rocks carried by fast-flowing water rub against each other. This action gradually wears away and smoothes the outer surfaces of the rocks (**Figure 3**). The force of pounding waves on a seashore can break large rocks into smaller fragments.

One of the most dramatic causes of mechanical weathering is glaciers. Although glaciers seem to stay in one place, they actually move very slowly downhill due to their immense weight. As they move, the rocks that are trapped in the ice scrape the ground below (**Figure 4**). This type of mechanical weathering is easily identified by the long scratches, called striations, it leaves on rocks.

Figure 4

During the last ice age, a huge glacier covered most of North America. You may find a rock like this, with marks that have been cut into it by rocks in the moving ice.

TRY THIS: *MODEL MECHANICAL WEATHERING*

Skills Focus: observing, inferring, creating models

Make a model to show how glaciers carrying rocks cause mechanical weathering. Fill an ice-cube tray with water. Sprinkle some sand in half of the sections. Freeze the water. Pop out a regular ice cube, and rub it along on a piece of foil. Then do the same with a sand cube. As you rub the sand cube along the foil, what do you observe (**Figure 5**)?

Figure 5

Chemical Weathering

As you have learned, rocks are made up of many different materials. Chemicals can weaken and break down some of these materials. **Chemical weathering** occurs when there is a chemical reaction between water, air, or another substance and the materials in rocks.

Water can dissolve some rock materials. If the water contains natural or human-made acids, the dissolving process will occur much more quickly. In nature, carbon dioxide gas in the air dissolves in rainwater to form a weak acid. When the rainwater passes over or through limestone, it dissolves some of the rock. Holes form in the rock. Over very long periods of time, these holes grow larger to form caves (**Figure 6**). In British Columbia, there are caves that were formed like this on Vancouver Island, in Glacier National Park, and near Mount Robson.

LEARNING TIP ◁

Check your understanding of how this cave formed by explaining it in your own words to a partner.

Figure 6
Rainwater dissolves the mineral calcite in limestone, sometimes forming large underground tunnels. The cave above is in Fernie, in the Southern Rockies of British Columbia.

Pollutants in the air can create acid precipitation (either rain or snow). Acid rain dissolves more minerals than normal rain. The limestone that is used to make statues and other monuments and buildings can be severely damaged by acid rain (**Figure 7**). Over many years, the mineral calcite in the limestone dissolves as acid rain pours over it, causing the limestone to crumble.

Air can also cause chemical weathering. The oxygen in air can rust any iron in minerals found in rocks.

Figure 7
The chemical weathering of ancient statues has been speeded up by the modern pollutants in acid rain.

TRY THIS: *MODEL CHEMICAL WEATHERING*

Skills Focus: creating models, observing, inferring

Create a model of chemical weathering by acid rain using chalk, water, and vinegar. Chalk is made of a compound called calcium carbonate. Calcium carbonate is found in the marble and limestone that are often used to make statues and buildings. Use a model to compare the effects of normal rain and acid rain on chalk. Put one piece of chalk in a glass or cup of tap water (rain). Put another piece of chalk in a glass or cup of vinegar (acid rain). Label the cups and leave the cups overnight. The next day, observe and compare the two pieces of chalk.

Biological Weathering

Sometimes living things cause mechanical or chemical weathering. This is called **biological weathering.**

Lichen grows on rocks and uses some of the materials in the rocks as a source of nutrients (**Figure 8**). It produces an acid that dissolves and wears down the rocks. When the lichen dies, it leaves a thin layer of weathered rock materials in which other plants can grow.

Figure 8
Lichen can wear down rocks.

Plants that grow in the cracks in rocks help to split the rocks. Wind and water deposit soil particles in the cracks caused by ice wedging. Roots grow in the cracks, splitting the rocks even more (**Figure 9**).

a) Trees grow very slowly where there is little soil. As the tree's roots grow, they split the rock.

b) This small tree may be hundreds of years old.

Figure 9
The roots of trees and other plants cause biological weathering.

TRY THIS: *OBSERVE BIOLOGICAL WEATHERING*

Skills Focus: observing, classifying

Take a walk around your schoolyard. Find and sketch examples of biological weathering. Look for lichen on rocks and weeds growing in cracks in the sidewalk. Post sketches in your classroom. Did other students find similar examples of biological weathering? Different examples?

▥▸ CHECK YOUR *UNDERSTANDING*

1. Draw a Venn diagram to show the relationships among mechanical, chemical, and biological weathering. On your diagram, define each type of weathering and give examples.

2. Explain how water can be involved in both mechanical and chemical weathering.

3. Old gravestones are sometimes so weathered that the writing is worn away. What types of weathering could act on a gravestone?

4. What human activities can increase the rate of weathering?

LEARNING TIP ◁

To review how to use a Venn diagram, see "Using Graphic Organizers" in the Skills Handbook.

7.6 Erosion

The movement of weathered rock materials from one place to another is called **erosion.** Erosion may occur rapidly, such as when a landslide races down a mountain. It may also occur slowly, over hundreds or thousands of years. Erosion can be the movement of grains of sand or the movement of gigantic boulders. Erosion can drop materials at any distance from their source—from a few centimetres to hundreds of kilometres away. When eroded rock materials stop moving, they settle on Earth's surface. The laying down of sediments is called **deposition.** Gravity, wind, water, and ice all help to move weathered rock materials.

Gravity

▷ **LEARNING TIP**

Scan the subheadings in this section. How many types of erosion do you think you will learn about?

Gravity causes rock falls and avalanches along many of British Columbia's highways. Early in the morning on January 9, 1965, a landslide near Hope, British Columbia, sent millions of tonnes of rock down a mountain into the valley below (**Figure 1**). Four people died, buried under rocks that reached depths of over 60 m. A small earthquake may have loosened the rocks, but the force of gravity caused the rocks to fall.

Figure 1
This rock slide, east of Hope, British Columbia, raced down the mountain with deadly force on January 9, 1965.

Wind

Wind can carry dust, sand, and soil for many kilometres. Particles in the air are deposited when the wind speed drops. On some beaches and deserts, the wind picks up dry, loose sand and deposits it in regular piles called dunes.

The devastating dust storms in the Prairies during the severe droughts of the 1930s demonstrated the wind's power (**Figure 2**). Due to the lack of rain, the rich surface soil became very dry. The wind was able to pick up the light, dry soil particles and blow them several kilometres away. In many places, the surface soil was completely blown away. The layers of soil that remained were not rich enough to grow crops, and many farmers were forced to abandon their farms. The effects of dust storms are not just local. Wind can even carry dust across oceans (**Figure 3**).

Figure 2
Dust storms on the Prairies in the 1930s damaged farmland.

Figure 3
The dark streak in the cloud approaching North America is dust from a dust storm in China six days earlier.

Water

Little by little, a large river like the Fraser River can move billions of tonnes of rock from the land it crosses. As the river does this, it cuts into the land and makes a deeper and deeper valley. A **valley** is any low region of land between hills or mountains. Valleys that are formed by flowing water tend to be V-shaped. On the way to the sea, many rivers cross flat areas, or **plains,** near the coast. Since a river moves slowly on a plain, the heavier sediment is deposited on the riverbed or riverbanks in the plains.

A river also slows down when it runs into a lake or ocean. Much of the sediment that the river was carrying is deposited on the bottom of the lake or ocean. As the sediment builds up, it causes the river to fan out over a broad area, often shaped like a large triangle. This area is called a **delta** (**Figure 4**). At a delta, the river often breaks into a number of smaller channels, separated by islands of sediment.

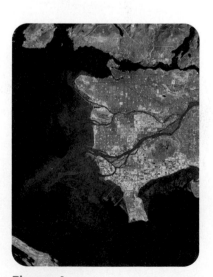

Figure 4
This satellite photograph shows the Fraser River delta in southern British Columbia.

The Fraser Canyon was carved out of the land by the action of flowing water (**Figure 5**). Over the years, the Fraser River cut deeper and deeper into the land until the canyon was formed. The Fraser River continues to carry rocks, gravel, sand, and mud. As time goes by, a large river extends its delta farther and farther out into the ocean. For example, 10 000 years ago, the end of the Fraser River delta was in the New Westminster area, 30 km east of where it is today.

Figure 5
The Fraser Canyon was carved out of the land by the action of flowing water.

Ice

Glaciers form when snow builds up over many years in the valleys and hollows of mountains (**Figure 6**). As the snow builds up, the layers are pressed together. The pressure gradually turns the snow to ice. The weight of the ice mass causes the glacier to move slowly downhill.

Figure 6
Bear Glacier, near Stewart, British Columbia, is a popular tourist attraction.

Rocks and soil that are frozen in the ice travel with the glacier. These materials can be carried many kilometres before being dropped in a new location when the glacier begins to melt. Glaciers can leave large, individual boulders in areas where they would not normally occur. These out-of-place boulders are called erratics (**Figure 7**).

Evidence indicates that the last ice age ended in most parts of Canada about 10 000 years ago. The effects of this ice age, however, can still be seen in many areas today. As rivers of ice moved down mountainsides, they eroded deep U-shaped valleys with round bottoms and steep sides. Along the coast, many of these valleys became filled with seawater after the glaciers melted. The resulting long, narrow inlets of the sea are called **fiords** (**Figure 8**). Howe Sound and Knight Inlet are two of the many fiords that are found along the coastline of British Columbia.

Figure 7
An erratic boulder.

Figure 8
Rivers Inlet is a fiord on the central coast of British Columbia.

⫸ CHECK YOUR UNDERSTANDING ⊗

1. What is erosion? How is it different from weathering?
2. List four forces that cause erosion, and give an example of each.
3. Give two examples of erosion that can happen quickly. Give two examples of erosion that happens slowly.
4. Deltas are often good areas for farming. Explain why.
5. How do you think Delta, in British Columbia, got its name?

Factors That Affect Erosion by Water

Moving water erodes soil (**Figure 1**). Several factors determine the amount of erosion that occurs and how fast it occurs.

> ▷ **LEARNING TIP**
>
> For help writing hypotheses or writing up your lab report, see the Skills Handbook sections "Hypothesizing" and "Writing a Lab Report."

Figure 1
Why might more erosion occur in some situations?

Question

What factors affect the rate of erosion by water?

Hypothesis

With your group, brainstorm a list of factors that could affect the rate at which soil is eroded by water. Write down each factor as a hypothesis. Then decide, as a group, which hypothesis you will investigate.

apron

plastic tray

pail of soil

container for water

Materials

- apron
- plastic tray
- 4-L pail of soil
- container for water

Decide if there are other materials you will need. Check with your teacher to make sure that these materials are safe for you to use.

- Design a procedure to test your hypothesis. A procedure is a step-by-step description of how you will conduct your experiment. It must be clear enough for someone else to follow and do the exact same experiment.

- Submit your procedure, including any safety precautions, to your teacher for approval. Also submit a diagram, at least half a page in size, showing how you will set up your experiment.

Data and Observations

Create a table to record your observations. Record your observations as you carry out your experiment.

Analysis

Compile the findings of all the groups in your class. Make a list of all the factors that affect the rate of erosion by water.

Conclusion

Look back at your hypothesis. Did your observations support, partly support, or not support your hypothesis? Write a conclusion for your investigation.

Applications

1. Where might you expect to see serious effects of water erosion?

2. How might erosion be a problem in your schoolyard or in your community?

3. Based on the findings of your class, how could you reduce the erosion of soil by water? Decide if there are other materials you will need to test your idea. Check with your teacher to make sure that these materials are safe for you to use.

▐▐▶ *CHECK YOUR* **UNDERSTANDING**   ⊗

1. Did any other groups test the same hypothesis that your group tested? If so, how were their results the same as, or different from, yours?

2. What advice about the design of your experiment would you give to another group of students testing the same hypothesis?

3. When designing your own experiment, why is it important to write down every step in the procedure?

The Rock Cycle

Over long periods of time, one kind of rock may change into another. The ways igneous, sedimentary, and metamorphic rocks change from one to another is called the **rock cycle** (Figure 1).

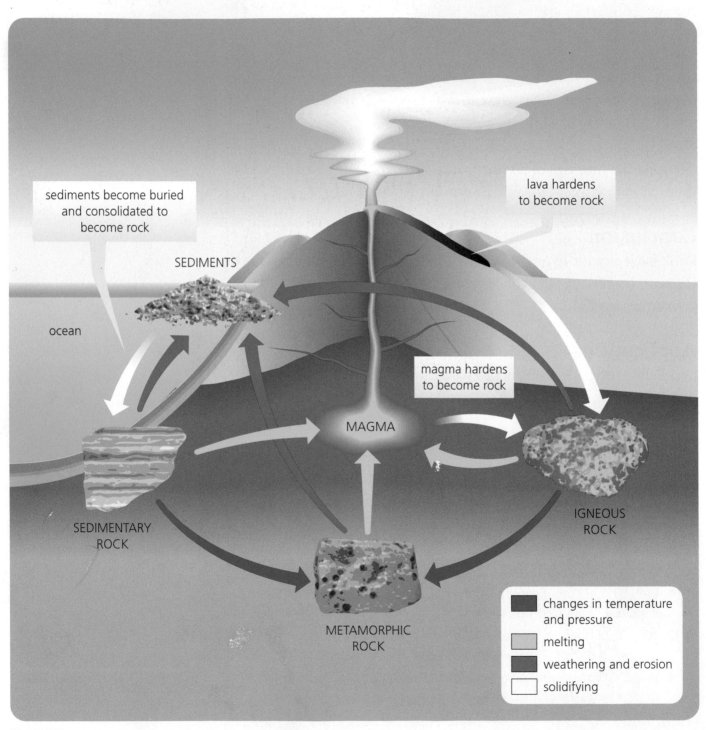

sediments become buried and consolidated to become rock

lava hardens to become rock

SEDIMENTS

ocean

magma hardens to become rock

MAGMA

SEDIMENTARY ROCK

IGNEOUS ROCK

METAMORPHIC ROCK

- changes in temperature and pressure
- melting
- weathering and erosion
- solidifying

Figure 1
The rock cycle

New igneous, sedimentary, and metamorphic rocks are constantly being formed. Rocks from all three families may eventually become exposed on Earth's surface, where weathering will wear them down. The resulting sediment will gradually form layers, which will get compressed into sedimentary rock.

Any rock, if pushed far enough into Earth, will become metamorphic due to the high temperatures and pressure beneath Earth's surface. If the rock is part of a continent or ocean floor and gets pushed deep within Earth, the rock will become extremely hot and melt, turning into magma. The magma can rise and erupt out of a volcano (**Figure 2**) or cool gradually near the surface, forming igneous rock. Each family of rock is linked to the others in this cycle.

Figure 2
This lava will harden to form igneous rock.

LEARNING TIP ◁
As you read the material on this page, follow the cycle in **Figure 1**. Look at the overall diagram, and then look closely at each part of the rock cycle. What family of rock is formed in each part of the cycle? What do the different coloured arrows mean?

▷ CHECK YOUR UNDERSTANDING ⊗

1. Why is the process in which rocks from one family change into rocks from a different family called the rock cycle?
   (*Hint:* What does "cycle" mean?)

2. How is the recycling of rocks by Earth like the recycling of newspapers, pop cans, or plastics by your community? How is it different?

3. Use the diagram of the rock cycle in **Figure 1** to explain how
   • an igneous rock can become a sedimentary rock
   • a sedimentary rock can become an igneous rock
   • an igneous rock can be broken down and become an igneous rock again

SKILLS MENU

○ Questioning	● Observing
○ Predicting	○ Measuring
○ Hypothesizing	○ Classifying
○ Designing Experiments	● Inferring
○ Controlling Variables	● Interpreting Data
● Creating Models	● Communicating

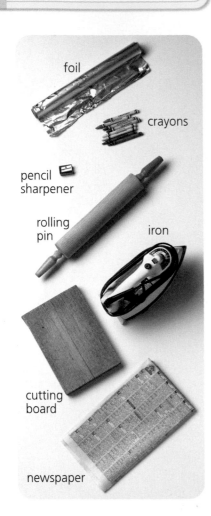

foil

crayons

pencil sharpener

rolling pin

iron

cutting board

newspaper

Creating a Model of the Rock Cycle

Creating a model allows you to observe something that is otherwise difficult to observe because it is too big, too dangerous, or takes too long. For example, you cannot observe some parts of the rock cycle because they take millions of years or happen deep within Earth. In this investigation, you will create a model of the rock cycle in your classroom.

Question

How can I model the changes to rock that take place in Earth?

Materials

- aluminum foil
- 8 crayons: 2 yellow, 2 grey, 2 green, and 2 purple
- pencil sharpener
- rolling pin
- iron
- heat-proof cutting board
- newspaper (optional; to protect work surface)

▐▶ Procedure

1 Remove the paper from around the crayons. Use the pencil sharpener to turn the crayons into shavings. Collect each colour of shavings in a separate pile.

2 On a 20 cm by 20 cm square of aluminum foil, carefully layer the crayon shavings by colour, the way that wind or water would deposit sediment.

3 Fold the aluminum foil over the layers of sediment. The foil package should be approximately 7 cm by 7 cm. Press the layers together with a rolling pin. Open the foil and sketch your "sedimentary rock."

4 Wrap your "sedimentary rock" in the aluminum foil again. Take the package to your teacher, who will press on the package with an iron set at medium-high heat for 5 to 8 s.

 Wax has a low melting and burning point. Let your teacher change your "sedimentary rock" into "metamorphic rock."

5 After your package has cooled, open it, and break your "rock" in half. How has it changed? Sketch your new "metamorphic rock."

Analyze and Evaluate

1. How did your "rock" change during the investigation? Use your sketches to help answer this question.

2. What did the rolling pin represent? What did the iron represent?

Apply and Extend

3. Based on your model, explain why sedimentary rock often appears in horizontal layers.

4. How could you extend your model to show igneous rock being formed?

▶ CHECK YOUR UNDERSTANDING ⊗

1. How did your model help you understand the formation of rocks?
2. How did your model differ from the real rock cycle?

Chapter Review

Old rocks can be recycled into new rocks.

Key Idea: Minerals are the building blocks of rocks and can be identified by their physical properties.

Vocabulary

minerals p. 179

Hardness

Colour, streak

Crystal structure, cleavage

Magnetism

Reaction with certain chemicals

Lustre

Key Idea: Earth's crust is made up of three families of rocks.

Vocabulary

magma p. 188

igneous rock
   p. 188

intrusive p. 188

lava p. 188

extrusive p. 188

sediment p. 190

sedimentary
   rock p. 190

metamorphic
   rock p. 192

Igneous

Sedimentary

Metamorphic

Key Idea: Fossils provide evidence of changes in life over time.

Vocabulary

fossils p. 194

fossil record
   p. 195

geologic time
   scale p. 196

Key Idea: Rock materials are broken down into smaller pieces by mechanical, chemical, and biological weathering.

Mechanical weathering

Chemical weathering

Biological weathering

Vocabulary

weathering p. 198

mechanical weathering p. 199

ice wedging p. 199

chemical weathering p. 201

biological weathering p. 202

Key Idea: Weathered materials are moved from one place to another by gravity, wind, water, and ice in a process called erosion.

Gravity

Water

Wind

Ice

Vocabulary

erosion p. 204

deposition p. 204

valley p. 205

plains p. 205

delta p. 205

fiords p. 207

Key Idea: Rocks and weathered rock materials can be transformed into new rocks.

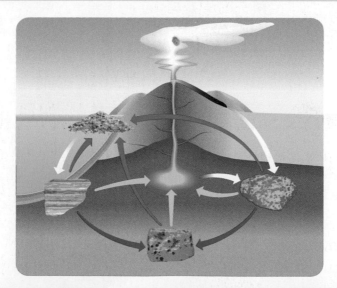

Vocabulary

rock cycle p. 210

Review Key Ideas and Vocabulary

When answering the questions, remember to use vocabulary from the chapter.

1. What properties of minerals are used to identify them?

2. Create a chart to show the three families of rocks and how each is formed.

3. How do fossils help geologists compare the age of rocks?

4. Make a table to explain the differences among mechanical, chemical, and biological weathering. Include examples of each type of weathering in your table.

4. Type of Weathering	Explanation	Examples
Mechanical		
Chemical		
Biological		

5. Explain how weathering is related to erosion.

6. Gravity, wind, water, and ice can all cause erosion. Give an example of how each can move weathered rock materials.

7. Use words and diagrams to explain how rock particles can go through the rock cycle from igneous to sedimentary to metamorphic rock, and back to igneous rock.

Use What You've Learned

8. Begin your own rock and mineral collection. A good way to start is to look for samples on weekends and during your holidays. You can also buy or trade samples through rock and mineral clubs and at shops that sell rocks and crystals. As well, you can look for scraps at building-stone suppliers.

9. Find out what minerals are mined in British Columbia. Report to the class on the minerals that are mined, where they are found, and their uses.

www·science·nelson·com **GO**

10. How can you tell whether a rock cliff beside a highway is made of igneous rock or sedimentary rock?

11. When learning new concepts it is sometimes helpful to think of something that you already know about. This is called an analogy. Look at the photographs below. Which type of cookie is an analogy for granite? Explain why.

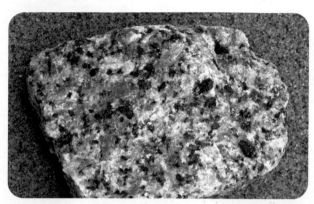

12. The Burgess Shale, near the town of Field, British Columbia, is one of the most famous fossil discoveries in the world. Research what types of fossils can be found there. Create a brochure that explains why these fossils are so important and why the Burgess Shale should be protected.

www·science·nelson·com

13. Which area would be more likely to lose its topsoil: a gently sloping area or a steep hill? Explain.

14. Name a natural or a human-made feature in your community that shows the effects of weathering and/or erosion. Make a diagram to show the kinds of weathering that are occurring. Has an attempt been made to prevent the weathering? If so, describe what has been done.

Think Critically

15. Look at a road map of British Columbia. What route would you take if you were travelling by car between Prince Rupert and Vancouver? Use the scale on the map to estimate the number of kilometres. About how many kilometres less would you travel if there were a highway straight down the coast? Why do you think there is not a more direct route between Prince Rupert and Vancouver?

16. List four or five local landforms. Predict how each landform might look in a million years.

Reflect on Your Learning

17. Recall what you have learned about rocks in this chapter. List some things that you did not know before you read this chapter. Then list any questions that you still have about rocks. Glance through the rest of the unit to see if your questions will be answered. If not, where can you go to find the answers?

18. Test yourself on your way home from school tonight. Look at different rocks and rock formations. See if you can identify them as igneous, sedimentary, or metamorphic rocks.

CHAPTER 8

Earth's crust is made up of moving plates.

⏩ **KEY IDEAS**

▶ Earth is made up of layers.

▶ Wegener developed his theory of continental drift using available evidence.

▶ Scientists now have additional evidence for the theory of plate tectonics.

▶ Earth's crust consists of slowly moving plates.

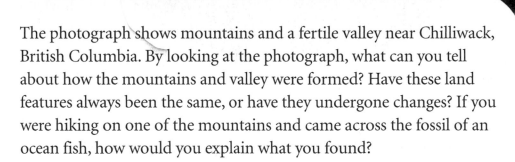

The photograph shows mountains and a fertile valley near Chilliwack, British Columbia. By looking at the photograph, what can you tell about how the mountains and valley were formed? Have these land features always been the same, or have they undergone changes? If you were hiking on one of the mountains and came across the fossil of an ocean fish, how would you explain what you found?

In Chapter 7, you learned that the surface of Earth is constantly being changed by weathering and erosion. You know that these slow processes are wearing away the mountains in the photograph. In this chapter, you will learn about the structure of Earth. You will learn about the slow, continuous processes that formed the mountains, and you will create models of Earth to show these processes. As well, you will interpret data supporting a scientific theory that was first proposed 100 years ago to explain these processes.

Earth: A Layered Planet

Imagine that you could drive a car at 100 km/h from the surface of Earth to its very centre. What would you see along the way? How long would your journey take? **Figure 1** shows what scientists think you would see.

LEARNING TIP ◁

As you read this section, use the arrows on the diagram to help you follow this journey to the centre of Earth. Look at the thermometers on the diagram to see how the temperature changes along the way.

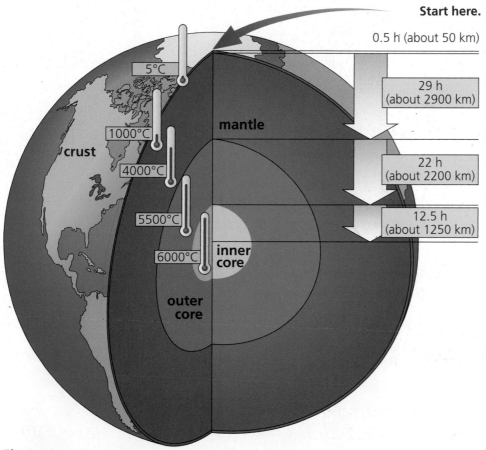

Figure 1
The layers of Earth

For the first half hour, you would pass through the crust. The **crust** is the thin layer of solid rock that makes up Earth's outermost layer. The materials in the crust tend to be lighter than the materials below. Earth's crust "floats" on the inner layers.

For the next 29 h, you would travel through the **mantle**, a hot, thick layer of solid and partly melted rock. Here you would begin to feel some pressure. The pressure would gradually increase because of all the layers above pushing down on you. The mantle moves sluggishly, like thick syrup.

Then you would travel for 22 h through the **outer core,** a dense, hot region that is made up mostly of liquid iron and some nickel. The pressure is very high. Like the mantle, the material in the outer core flows.

Finally, on the last part of your journey to Earth's centre, you would travel for 12.5 h through the **inner core,** a large ball of iron and nickel. Here, pressure from the weight of the other layers keeps the material solid, even though the temperature is almost as hot as the temperature on the surface of the Sun.

The idea of travelling to the centre of Earth is not new. In 1864, a French writer named Jules Verne published a novel called *Journey to the Centre of the Earth*. It describes the journey of a group of explorers as they try to reach Earth's core. In reality, the deepest hole that humans have made in Earth's crust is a mine that goes down approximately 12 km. At this depth, the temperature is already 70°C. The extreme heat and pressure at deeper levels prevent scientists from making a journey to the centre of Earth even today.

TRY THIS: MODEL A LAYERED EARTH

Skills Focus: creating models

Look at the photos of the orange, the peach, and the hard-boiled egg.

What are the strengths and weaknesses of each model of Earth? Which do you think is the best model? What layer of Earth does each part of this model represent?

▷ CHECK YOUR UNDERSTANDING

1. Draw a diagram of a cross-section of Earth and label the four layers.
2. How are the layers different from one another?
3. Your journey to the centre of Earth took 64 h at 100 km/h. What was the total distance you travelled?
4. Why have scientists never dug a hole to the centre of Earth?

Putting Together the Pieces of a Puzzle

In December 1910, a young German scientist named Alfred Wegener [VEG-nuhr] wrote in a letter to his girlfriend, "Doesn't the east coast of South America fit exactly against the west coast of Africa, as if they had once been joined? This is an idea I'll have to pursue." Two years later, Wegener presented his hypothesis that, a very long time ago, the continents were all part of one supercontinent he called **Pangaea** [pan-JEE-uh].

LEARNING TIP ◁

The word "Pangaea" comes from the Greek language and means "all lands."

The evidence that Wegener used to support his hypothesis came from the shapes of the continents and from fossils, landforms, and an ancient ice age. None of these observations were new. Other scientists had made these observations, but Wegener put them together and came up with a new scientific idea to explain them.

Shape of the Continents

Wegener observed that South America and Africa seemed to fit together like pieces of a jigsaw puzzle (**Figure 1**).

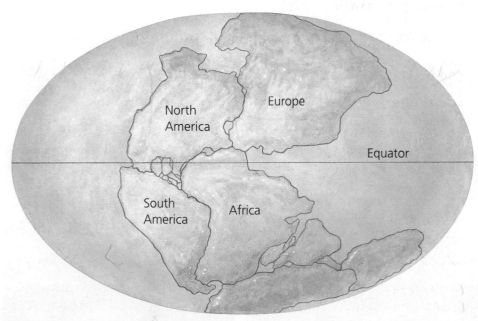

Figure 1
Were the continents once part of one giant supercontinent?

Fossil Record

Scientists had found fossils of identical plants and animals on different sides of the ocean (**Figure 2**). These plants and animals could not have travelled across the vast oceans, so they must have lived on the same continent at some time in the distant past. Scientists in Wegener's day hypothesized that there was once a land bridge between the continents, but Wegener disagreed. He thought that the continents had actually been joined.

▷ **LEARNING TIP**

When reading maps, remember to check the legend to find out what the different symbols or colours on a map represent.

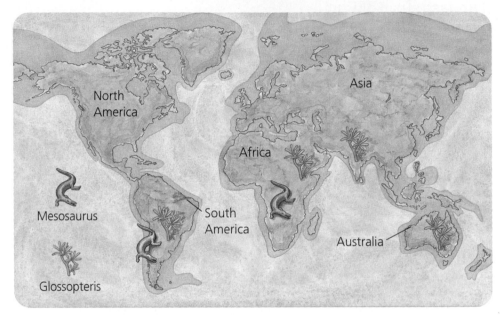

Figure 2

This map shows some of the fossil evidence Wegener used to support his hypothesis that the continents had once been joined to form Pangaea. Below are actual fossils of Mesosaurus (left) and Glossopteris (right).

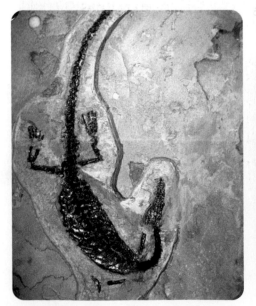
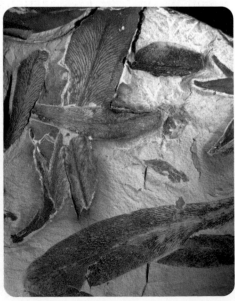

Landforms

Wegener noticed that when he put together the continents in his map of Pangaea, landforms on different continents matched. For example, mountains that run east to west across South Africa lined up with mountains in Argentina. Unusual rock formations and coal deposits in South Africa were the same as rock formations and coal deposits in Brazil. The Appalachian Mountains in the eastern United States matched the highlands of Scotland (**Figure 3**).

Ancient Ice Age

Scientists had found striations caused by ancient glaciers along the coasts of both South America and South Africa. The patterns formed by these striations were the same.

Scientists had also found deposits left by glaciers during an ancient ice age. Wegener found that on his map of Pangaea, the continents where this evidence had been found—Africa, India, Australia, and Antarctica—had once fit around the South Pole, where it would have been very cold.

Wegener said that, over time, the pieces of Pangaea separated, forming separate continents. He also said that the continents were still moving, or "drifting." He called this idea the theory of continental drift.

Imagine Wegener's disappointment when no one believed him. The main objection to Wegener's idea was that he could not come up with a good explanation for how the continents "drifted." Other scientists had difficulty imagining a way that huge continents could move thousands of kilometres.

◼ mountains
◼ coal

Figure 3
This map of Pangaea shows some of the landform evidence Wegener used to support his hypothesis that the continents had once been joined.

LEARNING TIP ◁

Always try to connect new information to things you have already learned. For example, think back to what you learned about mechanical weathering by glaciers. Are striations on the rocks caused by ice or by the rocks in the ice?

▸ CHECK YOUR UNDERSTANDING ⊗

1. What pieces of the puzzle did Wegener have? In other words, what evidence did he have to support his hypothesis that the continents had once been joined to form the supercontinent Pangaea?

2. What pieces of the puzzle did Wegener not have? In other words, what was the weakness in Wegener's theory of continental drift?

▷ **LEARNING TIP**

For a review about creating models, see the Skills Handbook section "Creating Models."

Creating a Model of Pangaea

Imagine that you are Alfred Wegener in the early 1900s. You are excited about your new hypothesis that today's separate continents were once joined together as Pangaea. You have evidence from the observations of other scientists, but the other scientists have different explanations for their observations. You are having a lot of trouble convincing them that your hypothesis about Pangaea could be correct.

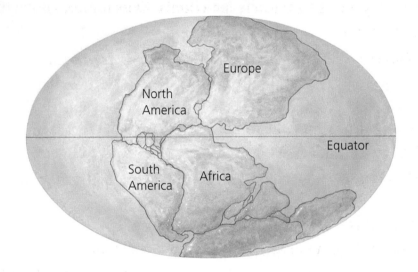

Problem

You want to make a model that will summarize all the evidence for Pangaea. You do not have the luxury of modern technology. You must use very simple materials to make your model.

Task

Using only the simple materials provided, develop a model of Pangaea. Your model must show Wegener's evidence for Pangaea, including evidence from the shapes of the continents, the fossil record, landforms, and an ancient ice age.

Criteria

To be successful, your model must

- be an accurate model of Pangaea
- include Alfred Wegener's evidence for Pangaea
- be made with only the materials listed

Plan and Test

Materials

- modelling clay in various colours
- paper or thin cardboard
- coloured pencils or markers
- dinner knife

modelling clay

paper

coloured pencils

dinner knife

Handle the knife carefully. Always cut away from yourself and others when using a knife.

Procedure

1. Using the materials listed above, make a model that shows Wegener's evidence for Pangaea.

2. Check your model against the criteria. Does it meet all the criteria?

3. If your model does not meet all the criteria, try again.

Evaluate

1. How does your model work? Does it show all of Wegener's evidence?

2. How is your model like the real Earth? How is it different?

Communicate

3. Draw a diagram to show how your model works.

4. What other materials could you use to make a model like this? List these materials, and explain how you would use them to represent moving continents. With your teacher's permission, create a new model. Compare your new model with your modelling-clay model. Which model is better? Why?

⫸ CHECK YOUR UNDERSTANDING ⊗

1. What are some of the limitations of your models?

2. Why are models useful?

3. Where else are models used to demonstrate ideas or represent real things?

8.4 Evidence for a New Theory

After Wegener died on an expedition to Greenland in 1930, his ideas were almost forgotten. It was not until the 1960s that new evidence made scientists reconsider his ideas about Pangaea and continental drift. By then, advances in technology had given scientists new information about the ocean floor.

Mapping the Ocean Floor

The ocean floor is not flat, as was previously believed (**Figure 1**). Scientists mapping the ocean floor were surprised to find deep trenches. These long, narrow trenches usually run parallel to and near the edges of the oceans. Scientists were also surprised to find a huge mountain range that almost encircles the Earth. This ridge of mountains is about 50 000 km long and runs through the middle of the oceans. It is called a **mid-ocean ridge.**

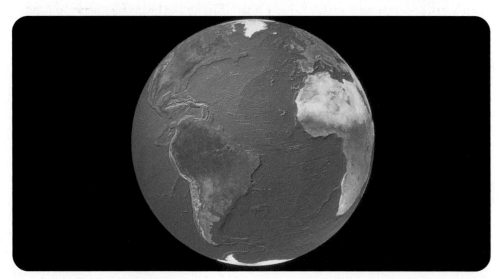

Figure 1
This map shows what the sea floor of the Atlantic Ocean would look like if all the water could be drained away. The ridge of mountains is called the mid-Atlantic ridge.

Ages of Rocks on the Ocean Floor

Scientists discovered that the layers of sediment on the ocean floor in the mid-Atlantic are quite thin. This suggested that the ocean floor is not as old as they had thought. If the ocean floor had remained unchanged for millions of years, the layers of sediment would have been thick.

As well, scientists found very young rocks at the top of mid-ocean ridges. The farther away from the ridge, the older the rocks are. Scientists concluded that the ridge is where the crust is splitting apart. Magma is rising at the ridge to form new crust. The sea floor at the mid-ocean ridge is increasing in size as new crust is formed. A scientist named Harry Hess called this process sea-floor spreading.

Hess suggested that the new ocean crust is constantly moving away from the ridge as if it is on a huge, and very slow, conveyor belt. After millions and millions of years, the crust sinks at a trench. Since Earth does not seem to be getting bigger, Hess concluded that the Atlantic Ocean is expanding and the Pacific Ocean is shrinking.

Magnetic Reversals

Scientists know that Earth's magnetic field has reversed several times over millions and millions of years. The mineral magnetite is magnetic. Grains of magnetite in molten magma line up like little magnets. The north ends of magnetite grains point to Earth's North Pole and the south ends point to Earth's South Pole. These patterns become locked into the rock as magma hardens. When Earth's magnetic field changes, this pattern reverses. Magnetite in solid rock cannot move, so only the magnetite in molten magma moves in the new direction.

The magnetic patterns locked into rock tell scientists about the direction of Earth's magnetic field at the time the rock was formed. Stripes of rock, parallel to the mid-ocean ridge, alternate between normal and reversed magnetic fields. This indicates that new rock is formed at the ridge.

TRY THIS: CREATE A MODEL OF A MID-OCEAN RIDGE

Skills Focus: creating models, observing, inferring

Push together two desks or tables. Take two pieces of lined paper. Hold the pieces of paper together beneath the desks, and push them slowly up through the crack as shown in **Figure 2**, about 4 cm at a time. Each time you stop, use a different colour to draw a stripe of rock on each side. Also draw arrows on each side to show the direction of magnetism. Reverse the direction of magnetism each time by reversing the direction of the arrow.

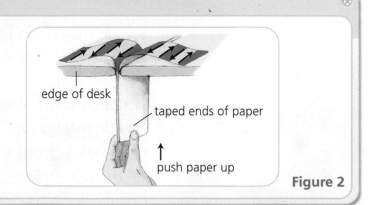

edge of desk

taped ends of paper

push paper up

Figure 2

Locations of Earthquakes and Volcanoes

Improvements in technology allowed scientists to record earthquakes and volcanoes under the oceans as well as on land. These new observations gave scientists a clearer pattern from which to draw conclusions about the movement of the continents and about Earth's crust in general. (You will learn more about the locations of earthquakes and volcanoes in Investigation 8.5.)

From these new observations, and the previous observations made by Wegener, the theory of **plate tectonics** [tek-TON-iks] was developed. According to this theory, the surface of Earth consists of about a dozen large plates that are continually moving (**Figure 3**).

> **LEARNING TIP**
>
> The term "tectonic" is used to refer to building or construction. It comes from the Greek word *tektonikos,* meaning "carpenter." So "plate tectonics" means "built or constructed of plates."

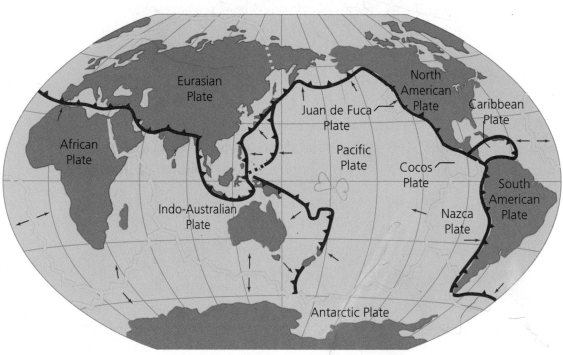

spreading apart

coming together (teeth on the side of the top plate)

direction of plate motion

Figure 3

The major plates of Earth's crust: One small plate that is very important to British Columbia is the Juan de Fuca Plate. This plate is sandwiched between the North American Plate and the Pacific Plate.

The parts of Earth's crust that have continents on them are called **continental crust.** The parts that have only ocean floor on them are called **oceanic crust.** Plates can be made up of continental crust, oceanic crust, or both. Wegener's theory of continental drift was wrong in one way: not only the continents are moving. Both the continents and the ocean floor form plates that move.

The plates move at different rates. The fastest plate movement, about 15 cm per year, is at the East Pacific Rise near Easter Island in the South Pacific Ocean. The Australian Plate moves about 6 cm per year. The slowest plate movement, about 2.5 cm per year, is at the Arctic Ridge.

Scientists are still working on the question that was such a problem for Wegener: What makes the plates move? Scientists generally agree with Hess's theory that the slow movement of the hot mantle below the plates moves the plates. However, the details about what causes this movement are still being discussed. Hess thought that the movement commonly seen in boiling water or soup played a role. Now scientists think that the sinking of oceanic crust into trenches, which pulls the rest of the plate behind it, is an important cause of plate motion. Unfortunately, none of the current theories fully explain all the observations about plate movement.

⫸ CHECK YOUR UNDERSTANDING

1. What is the theory of plate tectonics? How is it different from Wegener's theory of continental drift?

2. What new scientific evidence was added to Wegener's evidence to develop the theory of plate tectonics?

3. Why are we not able to observe with our senses that Earth's plates are moving?

4. In section 8.1, you used an orange, a peach, and a hard-boiled egg as models of Earth. Which one could you use to show plate movement? Explain how you would do this.

○ SKILLS MENU

○ Questioning   ○ Observing
● Predicting   ○ Measuring
● Hypothesizing ○ Classifying
○ Designing   ● Inferring
  Experiments
○ Controlling   ● Interpreting
  Variables   Data
○ Creating   ● Communicating
  Models

▷ **LEARNING TIP**

For help in writing hypotheses, see the Skills Handbook section "Hypothesizing."

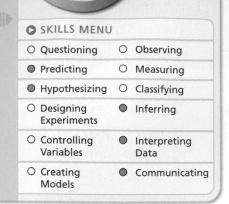

map

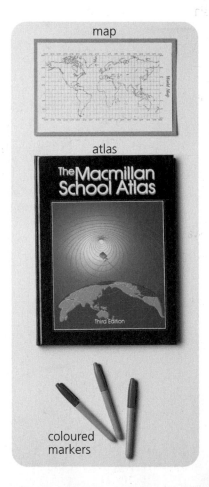

atlas

coloured markers

Finding Patterns in Geological Data

Scientists look for patterns in data to support theories. Geologists know that tremendous forces build up as tectonic plates move against each other. Is there a direct connection between plate movement and the location of mountain ranges, volcanoes, and earthquakes? If so, how would this connection support the theory of plate tectonics?

Question

Is there a pattern in the location of earthquakes, volcanoes, and mountain ranges (**Figure 1**)?

a) The 2003 earthquake in Bam, Iran, was devastating.

b) Scientists were able to warn people before Mount St. Helens, in the state of Washington, erupted in 1980.

Figure 1
Some areas of our planet seem to be more dangerous than others.

Hypothesis

Based on your knowledge of plate tectonics, make a prediction to answer each of the following questions:

(i) Where are earthquakes most likely to occur?
(ii) Where are volcanoes most likely to occur?
(iii) Where are mountain ranges most likely to occur?

Use your predictions to develop a hypothesis.

Materials

- map of the world with lines of longitude and latitude
- atlas or globe
- coloured markers

Procedure

1 Practise finding latitude and longitude on a map.
- To find latitude, measure from 0° at the equator up to 90° north (at the North Pole) or down to 90° south (at the South Pole).
- To find longitude, measure from 0° at the prime meridian (at Greenwich, England) either east or west to 180° at the International Date Line.
- As an example, look up the longitude and latitude of your home town or city and plot it on your world map.

2 On your world map, place a small blue circle ● at the location of each earthquake listed in **Table 1**.

3 Place a small red triangle ▲ on your world map at the location of each volcano listed in **Table 2**.

4 Use a third colour ■ to mark the locations of the following mountain ranges: Rockies, Andes, Himalayas, Alps, Urals, and Appalachians. You can use an atlas or a globe to help you find these mountain ranges.

Table 1 Major Earthquakes

Year	Location	Coordinates
1906	San Francisco, California	38°N, 122°W
1923	Tokyo, Japan	36°N, 140°E
1935	Quetta, Pakistan	30°N, 67°E
1939	Concepcion, Chile	37°S, 73°W
1964	Anchorage, Alaska	60°N, 150°W
1970	Yungay, Peru	9°S, 72°W
1972	Managua, Nicaragua	12°N, 86°W
1976	Guatemala City, Guatemala	14°N, 91°W
1976	Tangshan, China	40°N, 119°E
1985	Mexico City, Mexico	19°N, 99°W
1988	Shirokamud, Armenia	41°N, 44°E
1989	San Francisco Bay, California	38°N, 122°W
1990	Rasht, Iran	37°N, 49°E
1991	Valla de la Estrella, Costa Rica	10°N, 84°W
1993	Maharashtra, India	23°N, 75°E
1994	Northridge, California	34°N, 119°W
1995	Kobe, Japan	34°N, 135°E
1999	Taichung, Taiwan	24°N, 120°E
1999	Izmit, Turkey	41°N, 30°E
2001	Gujarat State, India	23°N, 70°E
2002	Hindu Kush Region, Afghanistan	36°N, 69°E
2003	Bam, Iran	29°N, 58°E

Table 2 Some Active Volcanoes

Volcano and Location	Coordinates
Etna, Italy	37°N, 15°E
Tambora, Indonesia	8°S, 117°E
Krakatoa, Indonesia	6°S, 105°E
Peleé, Martinique	14°N, 61°W
Vesuvius, Italy	41°N, 14°E
Lassen, California	40°N, 121°W
Mauna Loa, Hawaii	21°N, 157°W
Paricutin, Mexico	19°N, 103°W
Surtsey, Iceland	63°N, 20°W
Kelud, Indonesia	8°S, 112°E
Arenal, Costa Rica	10°N, 84°W
Eldfell, Iceland	65°N, 23°W
Mount St. Helens, Washington	46°N, 122°W
Laki-Fogrufjoll, Iceland	64°N, 18°W
Kilauea, Hawaii	22°N, 159°W
Mount Katmai, Alaska	58°N, 155°W
Avachinsky, Russia	53°N, 159°E
El Chichon, Mexico	17°N, 93°W
Ubinas, Peru	16°S, 71°W
Villarica, Chile	39°S, 72°W
Asama, Japan	36°N, 138°E
Shikotsu, Japan	41°N, 141°E

Analyze

1. Compare your map with the plate boundaries map. Describe any patterns you see in the locations of
   - earthquakes
   - volcanoes
   - mountain ranges

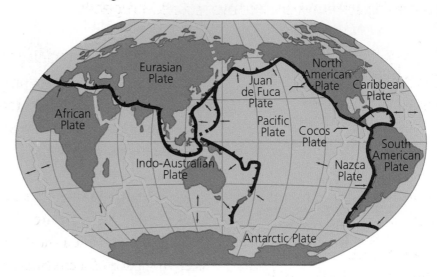

spreading apart

coming together
(teeth on the side of the top plate)

direction of plate motion

2. Are volcanoes always located near mountain ranges?

Write a Conclusion

3. Go back to your hypothesis. Do the patterns you found support, partly support, or not support your hypothesis? Write a conclusion for the investigation.

Apply and Extend

4. How do the patterns you found support the theory of plate tectonics?

5. Explain, using evidence from your map, why the edge of the Pacific Ocean is often called the Ring of Fire.

6. Which area of British Columbia is most likely to experience an earthquake? Explain.

▥▸ CHECK YOUR UNDERSTANDING

1. Was your hypothesis correct? If not, explain why not.

2. Maps can be important tools when looking for patterns. Think of another situation in which maps could be important for providing evidence.

Using Technology to Study Plate Movement

Scientists who study plate movement today have several technologies that were not available to Wegener.

Deep-Sea Submersibles

Deep-sea submersibles (**Figure 1**) allow scientists to investigate activity along mid-ocean ridges. Scientists now know that these are areas where plates are moving apart.

Alvin, used to explore the Mid-Atlantic Ridge, could operate at a depth of 4000 m. A submersible with a crew of two has reached the deepest part of the ocean, the 10 920 m Challenger Deep. Scientists do not yet have the technology to keep submersibles at that depth for the time it takes to explore the deep ocean.

Satellite Technologies

Satellites can be used to map features of Earth's surface and to measure distances between these features. The global positioning system (GPS) (**Figure 2**) allows scientists to make extremely accurate measurements. Even the creeping pace of a continent's movement is measurable by signals sent from satellites to the GPS receivers on Earth.

Fibre-Optic Technologies

Scientists are laying 3000 km of fibre-optic sensors along the border of the Juan de Fuca Plate and the North American Plate (**Figure 3**). This project, called the Neptune Project, will allow scientists at stations in Victoria, British Columbia, and in Oregon to collect information about even the smallest movement of these plates. The movement of these plates causes earthquakes in British Columbia.

Figure 1
The deep-sea submersible *Alvin*.

Figure 2
A temporary GPS receiver near the top of Mount Logan, British Columbia

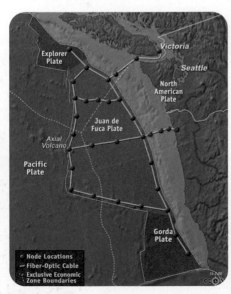

Figure 3
Thousands of kilometres of fibre-optic sensors will allow scientists to track the movement of the Juan de Fuca Plate.

Visit the Nelson Web site to learn more about the Alvin *submersible, GPS, and the Neptune Project.*

www·science·nelson·com (GO)

Earth's crust is made up of moving plates.

Key Idea: Earth is made up of four layers: the crust, the mantle, the outer core, and the inner core.

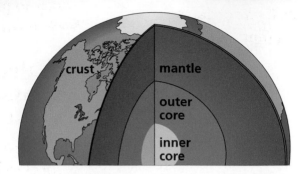

Vocabulary
crust p. 219
mantle p. 219
outer core p. 220
inner core p. 220

Key Idea: Wegener developed his theory of continental drift using available evidence—the shapes of the continents, the fossil record, landforms, and an ancient ice age.

Vocabulary
Pangaea p. 221

Key Idea: Scientists now have additional evidence for the theory of plate tectonics from

- mapping of the ocean floor
- the age of the rock on the ocean floor
- magnetic reversals
- the locations of earthquakes and volcanoes

Vocabulary
mid-ocean ridge p. 226
plate tectonics p. 228
continental crust p. 228
oceanic crust p. 228

Key Idea: Earth's crust consists of slowly moving plates.

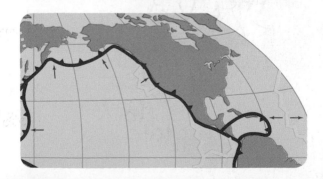

Review Key Ideas and Vocabulary

When answering the questions, remember to use vocabulary from the chapter.

1. Draw a diagram of the layers of Earth. Describe each layer in point form on your diagram.

2. Explain Alfred Wegener's theory of continental drift.

3. What evidence do scientists have to support the theory of plate tectonics?

4. How has technology helped scientists explore Earth?

Use What You've Learned

5. What is the name of your "home plate"—that is, the plate on which your home is located? Which plate or plates border on your home plate?

6. Look at the map of Earth's plates. Why do you think that many more volcanoes and earthquakes occur along the west coast of North America than along the east coast? (Refer to plates and the edges of plates in your answer.)

Think Critically

7. Why do you think it is important for scientists to learn more about Earth's plates?

8. What do you think is the most convincing piece of evidence for plate tectonics? Justify your answer.

Reflect on Your Learning

9. List two things you learned in this unit that you did not know before. What questions do you still have about the effects of plate movement?

10. Explain how learning about the theory of plate tectonics has changed the way that you think about
   (a) the area in which you live
   (b) Earth

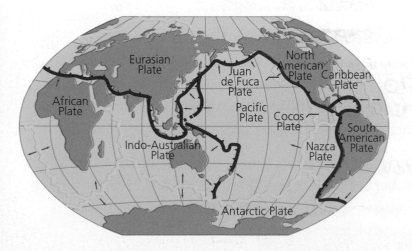

spreading apart

coming together
(teeth on the side of the top plate)

direction of plate motion

Plate movements cause both sudden and gradual changes to Earth's crust.

KEY IDEAS

▶ Moving plates interact at divergent, convergent, and transform fault boundaries.

▶ Earthquakes are caused by the sudden movement of plates at plate boundaries.

▶ Volcanoes can erupt at plate boundaries or at hot spots.

As you have learned, Earth's surface is constantly changing. Even though we may be unaware of it, mountains, valleys, and new islands are being formed. Some of the changes are sudden. Earthquakes can happen with little warning. Other changes are very gradual. The movement of the plates can create different kinds of mountains. Some changes, such as a volcanic eruption, are noticeable and newsworthy. Other changes, such as the formation of a ridge under the surface of the ocean, are seldom noticed or commented on. Even the slowest and least noticeable of these changes can have an enormous effect on the surface of Earth over a long period of time.

In this chapter, you will learn about both the sudden and gradual changes that alter the landscape of Earth. As you study this chapter, keep asking yourself what these changes have to do with the movement of the plates of Earth's crust.

Plates on the Move

9.1

Some of Earth's plates are being pulled apart, some are colliding, some are being pulled under others, and some are sliding past each other. The areas where plates meet are called plate boundaries. There are three types of plate boundaries:

1. divergent boundaries, where plates are being pulled apart

2. convergent boundaries, where plates are being pushed together

3. transform fault boundaries, where plates are sliding past each other

Divergent Boundaries

The boundaries between plates that are moving apart are called **divergent boundaries.** As plates separate, hot molten magma rises to Earth's surface to form new crust. On the ocean floor this separation of plates and production of new crust is called sea-floor spreading (**Figure 1**).

LEARNING TIP ◁

Preview this section and look at the headings. There are three main headings, one for each type of plate movement. Under one of the headings are three subheadings. Use this structure for taking point-form notes as you read the section. Your notes should always answer two questions:
- What is happening at this type of boundary?
- Where in the world is an example of this type of boundary?

Figure 1
This photo of the sea floor was taken by the deep-sea submersible seen at the bottom of the photograph. As plates move apart and create a crack in the crust, seawater seeps into the crack and becomes heated by the magma that is being pushed up from below.

The magma cools and hardens, forming ridges of new rock. These ridges can rise a kilometre above the ocean floor. The entire length of the Atlantic Ocean has a ridge in the middle where the North American and Eurasian plates are separating. This is known as the Mid-Atlantic Ridge (**Figure 2**). The rate of sea-floor spreading along the Mid-Atlantic Ridge is about 2.5 cm per year. Although this seems slow, it is gradually widening the Atlantic Ocean.

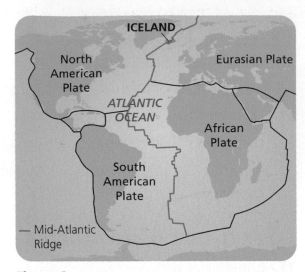

Figure 2

The Mid-Atlantic Ridge marks where the North American and Eurasian plates are separating.

On land, divergent boundaries create valleys called rifts. In Iceland, the divergent boundary between the North American Plate and the Eurasian Plate is visible on land (**Figure 3**).

Figure 3

A divergent boundary between separating plates is clearly visible as a rift near Thingvellir, Iceland.

Another divergent boundary is found in East Africa. Here the spreading between the African Plate and the Arabian Plate has already separated Saudi Arabia from the rest of the African continent and formed the Red Sea (**Figure 4**). It has also created the East African Rift system on land.

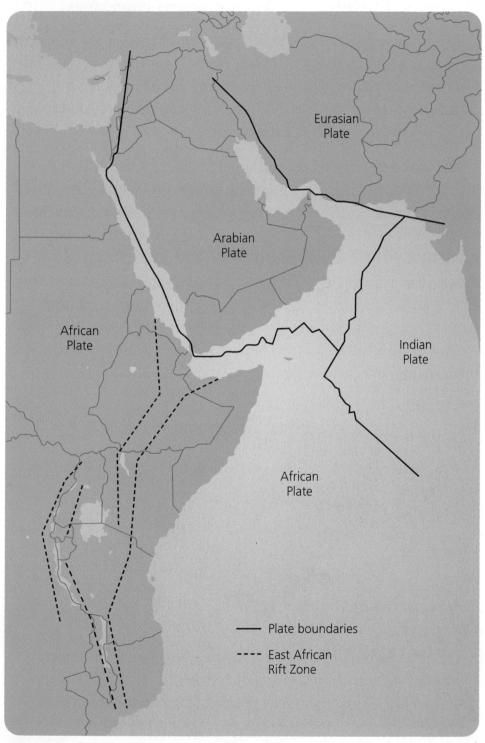

Figure 4
The divergent boundaries between plates in East Africa

Convergent Boundaries

Since there is no evidence that the size of Earth has changed significantly, old crust must be destroyed, or recycled, at the same rate that new crust is being formed at divergent boundaries. The recycling of old crust takes place at boundaries where plates move toward each other. These boundaries are called **convergent boundaries.**

The collisions that occur when plates come together are very slow and can last millions of years. When plates come together, one plate sinks below the other. The place where this occurs is called a **subduction zone** [sub-DUC-shun]. You can think of subduction as nature's way of recycling Earth's crust.

The landforms created at a convergent boundary depend on whether an oceanic plate is converging with a continental plate, two oceanic plates are converging, or two continental plates are converging.

Oceanic Plate Converging with Continental Plate

When an oceanic plate collides with a continental plate, the oceanic plate is subducted under the continental plate (**Figure 5**). This creates deep ocean trenches along the edge of a continent. Along the coast of British Columbia, this happens where the Juan de Fuca Plate is being subducted under the North American Plate.

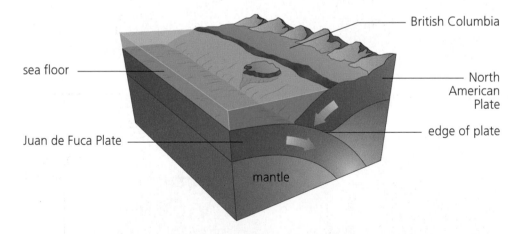

Figure 5
The Juan de Fuca Plate is being subducted under the North American Plate, which is moving west.

Off the coast of South America, the oceanic Nazca Plate is being subducted under the continental South American Plate. The South American Plate is being pushed up, creating the Andes Mountains. A **mountain** is any landmass that rises significantly from the surrounding level of Earth's surface. This type of mountain building is common where an oceanic plate is converging with a continental plate.

Oceanic Plate Converging with Oceanic Plate

When two oceanic plates converge, one plate sinks below the other. As with the convergence of oceanic and continental plates, trenches are formed at the subduction zone. Challenger Deep, the deepest part of the oceans, is part of the Mariana Trench in the subduction zone between the Pacific and Philippine Plates (**Figure 6**). Challenger Deep is so deep that even Mount Everest could not fill it in (**Figure 7**)!

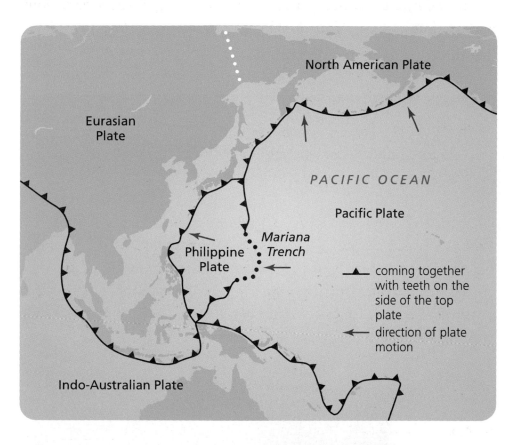

Figure 6
The Mariana Trench is being formed at a subduction zone.

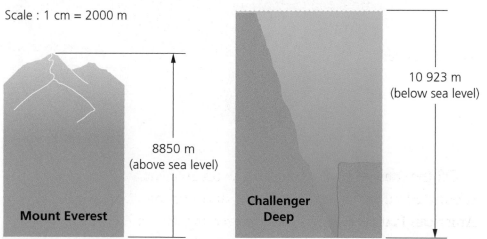

Scale : 1 cm = 2000 m

8850 m
(above sea level)

Mount Everest

10 923 m
(below sea level)

Challenger Deep

Figure 7
If you put Mount Everest into Challenger Deep, there would still be more than two kilometres of water over the top of Mount Everest!

LEARNING TIP ◁

Visualize turning Mount Everest upside down into Challenger Deep.

Continental Plate Converging with Continental Plate

When two continental plates meet, neither is subducted. Instead, the crust buckles and crumbles, pushing up mountains or areas of high level ground called **plateaus.** When the Indian Plate converged with the Eurasian Plate 50 million years ago, the slow uplift over millions of years pushed up the highest continental mountains in the world, the Himalayas (**Figure 8**). It also pushed up the Tibetan Plateau. Although the Tibetan Plateau is fairly flat, it is higher than the Alps mountain range in Europe.

Figure 8
The Himalayas

Transform Fault Boundaries

J. Tuzo Wilson, a Canadian geophysicist, made models of plate boundaries with paper and scissors. He discovered a new kind of plate boundary, which he called a fault. A fault is an area where rocks are being broken by movement in the crust. Wilson also discovered that divergent and convergent plate boundaries could end abruptly and "transform" into faults. He therefore called the zone between plates that are slipping past each other **transform fault boundaries.**

Most transform fault boundaries are found on the ocean floor. The most famous of the few on land is the San Andreas Fault (**Figure 9**) in California. At the San Andreas Fault, the Pacific Plate, which carries part of California, is moving north past the North American Plate, which carries the rest of California. Shallow earthquakes are very common along transform fault boundaries, such as the San Andreas Fault.

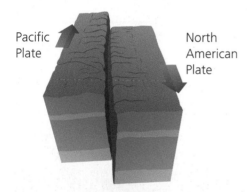

Pacific Plate

North American Plate

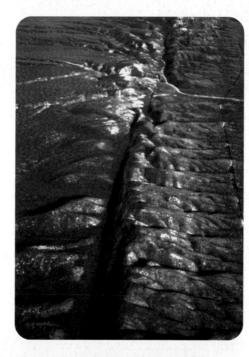

Figure 9
The San Andreas Fault is a transform fault boundary. Major earthquakes have often occurred along this fault.

⫸ CHECK YOUR UNDERSTANDING

1. Copy and complete **Table 1**.

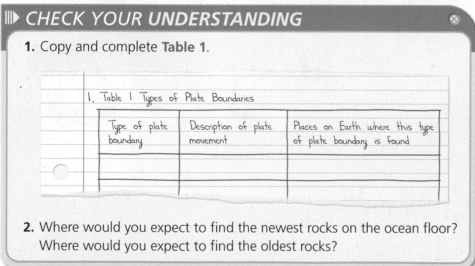

1. Table 1 Types of Plate Boundaries		
Type of plate boundary	Description of plate movement	Places on Earth where this type of plate boundary is found

2. Where would you expect to find the newest rocks on the ocean floor? Where would you expect to find the oldest rocks?

Creating Models of Plate Movements

The movement of plates and the changes to Earth's surface that result from this movement are too slow for you to observe directly. In this investigation, it will take you only a few minutes to use models to demonstrate the different ways that plates move over millions and millions of years.

Question

How can you demonstrate plate movement at the three types of plate boundaries?

Materials

- modelling clay in 4 different colours
- rolling pin
- dinner knife
- 2 sponges of different colours

 Handle the knife carefully. Always cut away from yourself and others when using a knife.

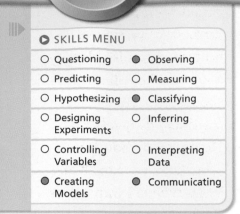

modelling clay

rolling pin

dinner knife

sponges

SKILLS MENU

○ Questioning ● Observing
○ Predicting ○ Measuring
○ Hypothesizing ● Classifying
○ Designing Experiments ○ Inferring
○ Controlling Variables ○ Interpreting Data
● Creating Models ● Communicating

Procedure

1 Use the rolling pin to flatten each colour of modelling clay to about 1 cm thick. Stack the four layers on top of each other and press down. Cut straight across the slab so that you have two equal pieces. These pieces will be one of your models for plates. The two sponges will be your other model.

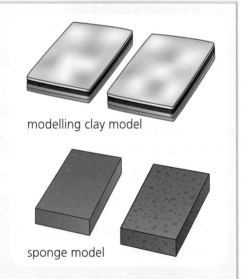

modelling clay model

sponge model

2 Using the sponge model, demonstrate plate movement for each type of plate boundary:
  - divergent
  - convergent
  - transform fault

Record your demonstration for each boundary with sketches.

3 Repeat step 2 using your modelling-clay model.

4 Use whichever model seems to work best for each of the following demonstrations:

- Demonstrate how movement at the boundary between the African Plate and the Arabian Plate separated Saudi Arabia from the rest of Africa. What type of boundary is this?

- Demonstrate how the Juan de Fuca plate is being subducted under the North American Plate. What type of boundary is this? Identify the type(s) of plates (oceanic and/or continental) involved.

- Demonstrate the formation of an ocean trench.

- Demonstrate how the Andes Mountains were formed. What type of boundary is this? Identify the type(s) of plates (oceanic and/or continental) involved.

- Demonstrate how the Himalayas were formed. What type of boundary is this? Identify the type(s) of plates (oceanic and/or continental) involved.

- Demonstrate how a landform (hill or stream) can be separated by plates sliding past each other along a transform fault boundary.

Analyze and Evaluate

1. Which model worked best for each demonstration? What were the limitations of the other model?

2. Were you able to create mountains in more than one way with colliding plates?

Apply and Extend

3. What were the limitations of your models in demonstrating the formation of a rift valley? How could you modify or add to one of your models to improve the demonstration?

4. What other materials could you use to model plate movement?

▶ CHECK YOUR UNDERSTANDING

1. What are the advantages and disadvantages of using models to show plate movement?

9.3 Earthquakes

▷ **LEARNING TIP**

Preview the headings on these two pages. What can you say about where earthquakes occur?

As the plates that make up Earth's crust move, the rough edges lock together. Over time, pressure builds up until one or both of the plates suddenly move, releasing the energy stored in rocks. The sudden release of energy causes vibrations of Earth's crust called **earthquakes.**

Earthquakes can occur at all three types of plate boundaries:

1. divergent
2. convergent
3. transform fault.

Earthquakes at Divergent Boundaries

Earthquakes can occur where two plates are being pushed apart. Hot magma rising below the crust pushes upward toward an opening in the crust (**Figure 1**). Pressure builds up where the plates are joined. Then suddenly, the pressure is enough to push the plates apart, and the crust shakes. This produces a small local earthquake. There are constant small earthquakes along the Mid-Atlantic Ridge.

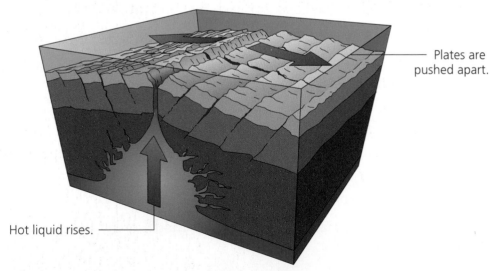

Plates are pushed apart.

Hot liquid rises.

Figure 1
Some earthquakes occur as two plates are pushed apart.

Earthquakes at Convergent Boundaries

When an oceanic plate is subducted under another oceanic plate or a continental plate, it may get stuck against the top plate (**Figure 2**). The force builds up until the top plate suddenly moves. This sudden movement can cause a large earthquake. The longer a plate is stuck, the stronger the earthquake is when the plate breaks free. Southern British Columbia experiences almost 200 earthquakes a year as the Juan de Fuca Plate is subducted under the North American Plate.

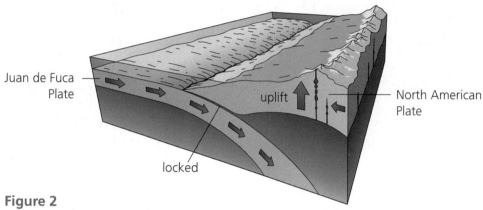

Figure 2
Some earthquakes occur in subduction zones.

Earthquakes at Transform Fault Boundaries

When two plates are moving past each other in opposite directions along a transform fault boundary, they sometimes get stuck (**Figure 3**). The force builds up until one plate suddenly moves, causing an earthquake. The longer the time before the plates slip, the stronger the earthquake is. There have been many powerful earthquakes of this type along the San Andreas Fault in California.

LEARNING TIP ◁

Check your understanding of why earthquakes occur at plate boundaries by explaining **Figures 1**, **2**, and **3** to a partner.

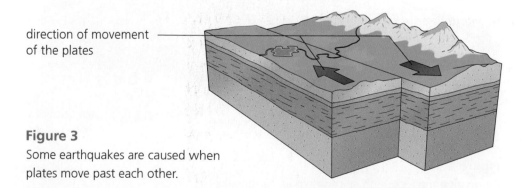

Figure 3
Some earthquakes are caused when plates move past each other.

The Effects of Earthquakes

News reports and newspaper articles about earthquakes usually include large, dramatic photos of the damage that earthquakes cause (**Figure 4**). How does this damage occur?

Figure 4
Earthquake damage in Mexico City in 1985.

The exact location within Earth at which an earthquake starts is called the focus (**Figure 5**). The focus is often deep within Earth's crust. The point on the surface of Earth directly above the focus is called the epicentre of the earthquake. The shock waves that are sent out when an earthquake occurs are called **seismic waves.** Smaller tremors can occur at any time for months after an earthquake as the pressure within Earth's crust is gradually released. These tremors are called **aftershocks.**

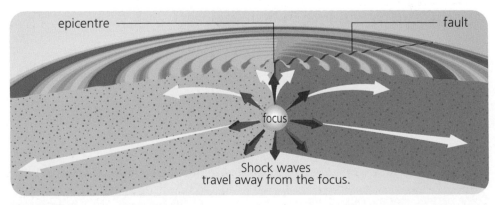

Figure 5
Comparing the focus and epicentre of an earthquake.

The energy that is released from the focus travels outward in all directions. The strength of the earthquake depends on the amount of energy that is released from the plate movement. There are two main types of seismic waves: primary (P) waves and secondary (S) waves. These waves and their effects are compared graphically in the Venn diagram in **Figure 6**.

Primary (P) Wave
- travels through liquids and solids
- pushes and pulls materials as they move through Earth
- travel about 8 km per second
- cause the first movement you feel in an earthquake

Both
- originate from same focus
- begin at same time
- can be felt at Earth's surface

Secondary (S) Wave
- travels through solids only
- makes the rocks vibrate up, down, or sideways
- travel at about 4.5 km per second
- usually cause more building damage

Figure 6
The two types of seismic waves that are produced by an earthquake cause different effects.

Geologists cannot observe Earth's mantle and core directly. They use indirect evidence from seismic waves to infer the characteristics of the interior of Earth (**Figure 7**).

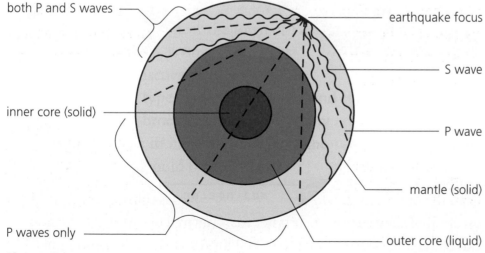

both P and S waves

earthquake focus

S wave

inner core (solid)

P wave

mantle (solid)

P waves only

outer core (liquid)

LEARNING TIP

Look at the overall diagram of earthquake waves travelling though Earth. Then look closely at each type of wave (P or S) separately and follow its path. If you are not sure why their paths are not the same, re-read the caption.

Figure 7
When an earthquake starts at the focus, the P waves can be detected anywhere. The S waves can be detected only at the locations shown. Since S waves cannot travel in liquid, scientists assume that part of Earth's interior must be liquid. This liquid part is called the outer core.

When an earthquake begins, the ground starts to shake, causing buildings to sway back and forth. If this happens in a rural area, only a few people may be in danger. If this happens in a city or town, many people may be affected. In addition to damaging buildings and roads, earthquakes can cause tunnels and overpasses to collapse. Fires can start when fuel tanks and gas lines break (**Figure 8**). Water lines can also break, leaving people without drinking water or water to fight fires.

Figure 8
In 1995, an earthquake caused massive damage when it hit the city of Kobe, Japan. Over 300 fires were started as a result.

▷ **LEARNING TIP**

The name "tsunami" comes from the Japanese words for harbour (津 or *tsu*) and wave (波 or *nami*). Why is this name a good choice?

If a large earthquake causes a section of the sea floor to move, a series of ocean waves is created. Ocean waves that are caused by an earthquake or an underwater volcano are called **tsunamis** [tsu-NAH-mees].

In the open ocean, tsunamis are small and pass almost unnoticed, except that they travel much faster than normal ocean swells. Tsunamis can travel at speeds of more than 800 km/h, as fast as a commercial jetliner, in deep water. When they approach shallower waters, their speed is reduced but they grow to surprising heights. In narrow inlets or shallow harbours, their height can increase to 30 m or more. Tsunamis can cause massive destruction and flooding in coastal areas.

Tsunamis from the Anchorage, Alaska, earthquake of 1964 hit the coast of British Columbia. The Hesquiaht [HESH-dwit] village at the head of Hot Spring Cove was completely wiped out (**Figure 9**). As well, there was considerable damage to Port Alberni because the narrow Alberni Inlet pushed the waves to greater heights.

Figure 9
In 1964, tsunamis moved nearly every house in this Hesquiaht village at Hot Spring Cove off its foundation, forcing the people to rebuild elsewhere.

CHECK YOUR UNDERSTANDING

1. What is the main cause of earthquakes?
2. List the three types of plate boundaries. Explain why earthquakes can occur at each type of boundary. Use sketches and diagrams in your explanation.
3. Which two types of plate movement produce the biggest earthquakes? Why do you think the other type of plate movement produces smaller earthquakes?
4. Why do seismologists (scientists who study earthquakes) worry if a plate stops moving?
5. What are aftershocks? Why do aftershocks present a special danger?
6. What are tsunamis? Describe how tsunamis are produced.

▷ **LEARNING TIP**

To review the steps in solving a problem or researching see the Skills Handbook sections "Solving a Problem" and "Researching."

Preparing for an Earthquake
Problem
Are you prepared for the big one? Since many parts of British Columbia are in an earthquake zone, it is only a matter of time before a major earthquake occurs. You need to be prepared for a major earthquake, both at school and at home. What can you do before, during, and after an earthquake to keep yourself as safe as possible?

Task
Create two Earthquake Preparedness Checklists—one for school and one for home.

Criteria
To be successful, your checklists must

- provide details about the dangers during and after an earthquake
- describe precautions that you can take before, during, and after an earthquake
- be thorough and accurate
- be clear and easy to understand

Plan and Test
1. Research dangers that you might expect during a major earthquake. What can you do now to prepare for these dangers? What can you do during an earthquake to keep yourself as safe as possible from these dangers? Are the dangers at home different from those at school? Take notes as you do your research.

2. Now research dangers that you might expect after an earthquake. What can you do now to prepare for these dangers? What can you do during and after an earthquake to keep yourself as safe as possible from these dangers? Are the dangers at home different from those at school? Take notes as you work.

www·science·nelson·com GO

3. Use your notes to draft school and home Earthquake Preparedness Checklists. Use the following headings: Before an earthquake, During an earthquake, After an earthquake.

Evaluate

4. Compare your draft school and home checklists with those of other students. Is there anything that you think you should add?

5. Compare your school checklist with the school's earthquake preparedness plan, if there is one. Is there anything that you think you should add to your checklist?

6. Show your home checklist to your parents. Is there anything that they think you should add or change?

Communicate

7. Make final copies of your school and home checklists. You may want to illustrate your checklists to emphasize important points.

8. Use your school checklist to teach a group of younger students about how to be prepared for an earthquake.

9. Is there anything on your checklist, or your classmates' checklists, that you think should be added to your school's plan? If so, get advice from your teacher about how to communicate this to your principal.

10. Is there anything on your checklist that your family has not done? Talk to your family about why it is important to do everything on your checklist.

▶ CHECK YOUR UNDERSTANDING ⊗

1. What resources did you use to create your checklists?

2. Are there organizations in your community that could provide you with additional information? Compare your checklists with any information that these organizations provide.

Indigenous Knowledge Sheds Light on British Columbia's Geological Past

Native carver Tim Paul depicts the earthquake god as a fearsome relative of humanity. While humanity's other 10 relatives, such as Moon and stars, nurture us, the earthquake god humbles us.

On a still midwinter night, long before Europeans first landed on Vancouver Island, native legend tells of a great disaster. The sea rose in a heaving wave, and landslides buried a sleeping village.

"They had practically no way or time to save themselves. They simply had no time to get hold of canoes, no time to get awake," the late Nuu-chah-nulth Chief Louie Clamhouse told Alan McMillan, an anthropologist at Simon Fraser University. "I think a big wave slammed into the beach. The Pachena Bay people were lost."

Over time, storytellers began to speak of dwarfs in the mountains, mythic creatures who would dance around their great wooden drum, causing Earth to shake and the waters to rise.

In 2003, government research proved that an earthquake, the most intense Canada has ever seen, hit the sea floor off the British Columbia coast at 9:00 P.M. on January 26, 1700. Earthquakes of that intensity cause tsunamis, and Japanese written history tells of a massive tsunami striking fishing villages the next day along the coast of Honshu, killing hundreds. Coupled with geological evidence of the level 9 quake, the connection was clear.

Tim Paul, a Nuu-chah-nulth carver and silkscreen artist, has recorded these earthquake legends in his art. In one celebrated mask, he casts the earthquake god Ta-gil as a terrifying, cave-dwelling "relative" of humanity. The other 10 relatives—Sun, stars, Moon, and so on—nurture us, but Ta-gil "puts us in our place" and with his destruction, he "reminds us that we are the smallest part of nature," Mr. Paul says.

In Mr. Paul's silkscreens, Ta-gil is depicted with an enlarged foot, the "earthquake foot," that enables him to shake the ground. The Juan de Fuca tectonic plate, whose motion is responsible for all these myths, is the smallest tectonic plate on Earth and thus the easiest to study.

One of the most spectacular shows in nature occurs when a volcano erupts (**Figure 1**). Any opening in Earth's crust through which molten rock and other materials erupt is called a **volcano.**

Volcanoes and earthquakes are proof that, deep within our planet, there are tremendous forces at work. Like earthquakes, most volcanoes are located along the edges of Earth's plates. Only a few volcanoes are found away from the edges of plates.

Some volcanoes erupt frequently and relatively quietly. You can actually watch the lava flow out of these volcanoes from a safe distance. Other volcanoes only erupt once every few hundred years, but with massive explosions. Many volcanoes go unnoticed at the bottom of oceans.

LEARNING TIP ◁

Make a three column K-W-L chart. Record what you already know about volcanoes in the first column. Record what you wonder about them in the second column. After you finish reading this section, write what you have learned in the third column.

Figure 1
Molten lava flows from a Hawaiian volcano into the ocean.

Volcanoes at Divergent Boundaries

When you hear the word "volcano," you probably think of a volcano on land. However, about three-quarters of all lava produced on Earth comes from eruptions at divergent boundaries on the ocean floor. Magma pushes to the surface where plates are moving apart (**Figure 2**). The lava erupts and cools to form a ridge on each side of the crack on the ocean floor. Some of these ridges may rise high enough to reach the surface, creating islands. Iceland was formed, and continues to be formed, in this way.

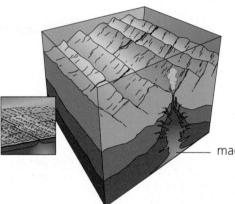

magma beneath the plate edge

Figure 2
The Mid-Atlantic Ridge was formed by magma pushing two plates apart.

In 1963, plumes of smoke billowed out of the ocean near Iceland (**Figure 3**). Soon a new island appeared, as large amounts of magma flowed out of the ocean rift between the Eurasian and North American Plates. Eventually, the island grew to 150 m above sea level.

Mount Edziza, Hoodoo Mountain, Lava Fork, and Crow Lagoon are volcanoes on a divergent boundary in the northwestern corner of British Columbia, near the border of Alaska. These are the youngest volcanoes in the province. It has been 150 years since one erupted.

Figure 3
The new island formed off the coast of Iceland was named Surtsey, after Surt, the lord of the land of fire giants in Norse mythology.

Volcanoes at Convergent Boundaries

Most of the volcanoes on land are located near convergent plate boundaries. Some of the most powerful volcanic eruptions occur where one plate is being subducted under another plate. The magma that is formed in a subduction zone is thick and sticky. Since the magma is too still to allow steam and volcanic gases to escape, tremendous pressure builds up. This type of volcano erupts explosively as the pressure is released. As the lava reaches the surface, the high-pressure steam escapes, carrying the lava and ash with it (**Figure 4**).

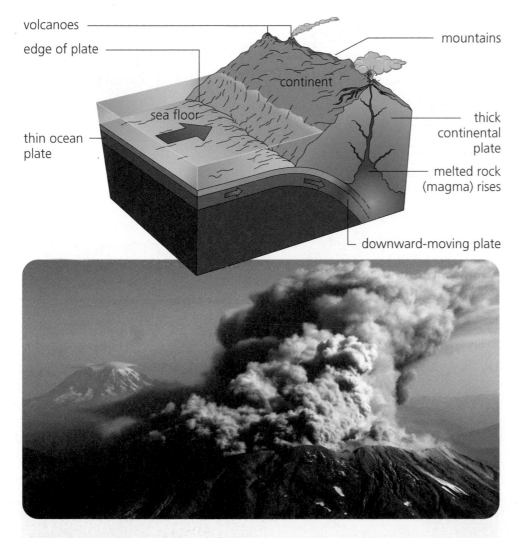

Figure 4
In 2004, Mount St. Helens, in the state of Washington, erupted. Mount St. Helens is near the convergent plate boundary where the Juan de Fuca Plate is being pulled under the North American Plate.

There has not been an explosive volcanic eruption in British Columbia since Mount Meager, near present-day Whistler, erupted over 2000 years ago. Mount Silverthrone, Mount Cayley, and Mount Garibaldi are other volcanoes on the convergent boundary where the Juan de Fuca plate is being subducted under the North American plate.

Volcanoes That Form at Hot Spots

Although most of Earth's volcanoes occur near plate boundaries, there are some exceptions. The Hawaiian Islands formed from volcanoes in the middle of the Pacific Ocean, over 3000 km from the nearest plate boundary. This puzzled scientists until J. Tuzo Wilson discovered evidence of hot spots. **Hot spots** are parts of the mantle where the temperature is much higher than normal.

At a hot spot, magma collects in enormous pools. The hot magma eventually melts a hole in the rock above it and pours out of the hole onto Earth's surface as lava (**Figure 5**). The lava that is produced at a hot spot tends to be runny and so fluid that the volcano does not erupt explosively. The lava simply pours out of the volcano like a river and hardens.

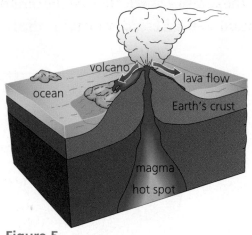

Figure 5

Volcanoes can form at a hot spot, where a huge pool of hot magma has risen through the mantle and melted a hole through the solid rock of Earth's crust.

If this type of volcano forms on the ocean floor, the lava hardens more quickly than it would on land. The hardened lava forms a cone-shaped mountain that may eventually rise above sea level as an island. This is how the Hawaiian Islands were formed (**Figure 6**).

Figure 6

Mauna Loa, on the island of Hawaii, is a volcano that formed at a hot spot. From its base at the bottom of the Pacific Ocean to its summit, it rises 9750 m. This makes it taller than Mount Everest!

Recently, an underwater volcanic peak rising above the ocean floor was discovered close to the southern coast of the island of Hawaii. Scientists call it Loihi. It is over 3000 m above the floor of the ocean, but it is not expected to become an island officially for another 45 000 years.

British Columbia has a small hot spot area called the Anahim Volcanic Belt. This area stretches from the coast to Quesnel. **Figure 7** summarizes the locations where all three types of volcanoes are found in British Columbia.

LEARNING TIP ◁

It is easier to remember information if you personalize it. How might a volcano affect your life?

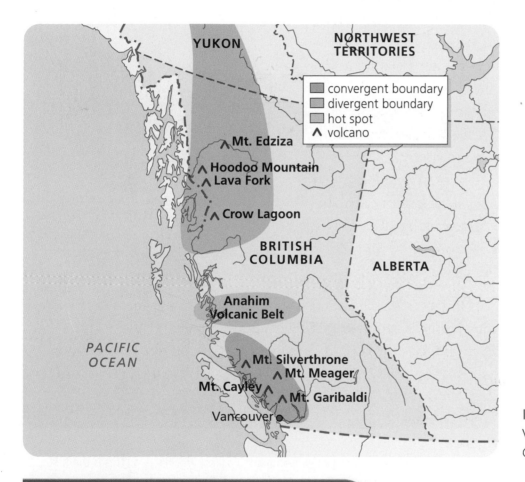

Figure 7
Volcano locations in British Columbia.

TRY THIS: CLASSIFY VOLCANOES

Skills Focus: inferring, classifying

Go back to the map you labelled in Investigation 8.5. Beside each volcano you labelled, indicate whether it is occurring near a

- divergent boundary
- convergent boundary
- hot spot

Design your own symbols to use on your map.

The Effects of Volcanoes

Earthquakes can be very destructive, killing people and destroying property. People have been killed by clouds of hot ash and poisonous gases, and buried by volcanic mudslides. People have died of starvation because their crops, livestock, or other sources of food were destroyed. Whole villages and even cities have been destroyed by volcanoes. About 250 years ago, a lava flow in northern British Columbia destroyed two villages and killed about 2500 people (**Figure 8**). Scientists believe they were killed by carbon dioxide gas.

▷ **LEARNING TIP**

Ask yourself, "How could community planners make use of this information on the effects of volcanoes?"

Figure 8
Lava beds at Anhluut'ukwsim Laxmihl Angwinga'asanskwhl Nisga'a: This is the first provincial park in British Columbia that has been established to combine the interpretation of geological features and Aboriginal culture.

Not all the effects of a volcano are felt right away or only close to the volcano. Ash from the 1980 eruption of Mount St. Helens fell on Vancouver. Large, explosive eruptions can send ash and gases high into the atmosphere. Volcanic clouds from the eruptions of Tambora (Indonesia) in 1815 and Mount Pinatubo (Philippines) in 1991 drifted around Earth, blocking the Sun and cooling temperatures for years.

Volcanoes have positive effects, as well. Volcanic ash improves soil and creates rich farmland. Volcanic rocks contain many useful minerals and gems. Some of the largest diamonds in the world are found in volcanic rocks. People have been using volcanic rocks for thousand of years. The Tahltan [TALL-tan] First Nation was mining obsidian several thousand years ago because the sharp edges and points of this volcanic rock could be made into useful tools (**Figure 9**). Some Aboriginal peoples used obsidian scalpels for surgery. This indigenous knowledge is important today— obsidian scalpels are still used for eye surgery.

Figure 9
Aboriginal people in many areas used the volcanic rock obsidian to make tools.

⫸ CHECK YOUR UNDERSTANDING

1. Copy and complete **Table 1** to summarize what you have learned about the three main types of volcanoes.

   Table 1 Types of Volcanoes

Location	Characteristics	Examples

2. People were killed when Mount St. Helens erupted, but tourists can watch volcanic eruptions in Hawaii. Why are some volcanoes more explosive than others?

3. The ancient Roman city of Pompeii was destroyed by an earthquake in 63 C.E. Pompeii was rebuilt, only to be destroyed again in 79 C.E. when a nearby volcano, Vesuvius, erupted and buried the city in ash. Why are many volcanoes found in the same areas that earthquakes occur?

4. All the volcanoes in Canada are located in British Columbia and the Yukon Territory. Explain why.

9 Chapter Review

Plate movements cause both sudden and gradual changes to Earth's crust.

Key Idea: Moving plates interact at three types of boundaries:

Divergent boundary

Convergent boundary

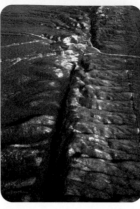
Transform fault boundary

Vocabulary

divergent
   boundaries p. 237

convergent
   boundaries
   p. 240

subduction zone
   p. 240

mountain p. 240

plateaus p. 242

transform fault
   boundaries
   p. 243

Key Idea: Earthquakes are caused by the sudden movement of plates at plate boundaries.

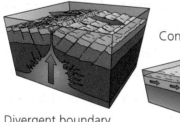
Divergent boundary

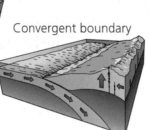
Convergent boundary

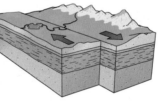
Transform fault boundary

Vocabulary

earthquakes
   p. 246

seismic waves
   p. 248

aftershocks
   p. 248

tsunamis p. 250

Key Idea: Volcanoes can erupt at plate boundaries or at hot spots.

Divergent boundary

Convergent boundary

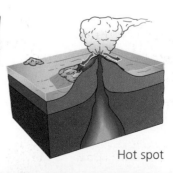
Hot spot

Vocabulary

volcano p. 255

hot spots p. 258

Review Key Ideas and Vocabulary

When answering the questions, remember to use vocabulary from the chapter.

1. What are the three types of plate boundaries? Sketch a diagram to explain plate movement at each.

2. What causes earthquakes? Where do earthquakes usually occur?

3. Where are volcanoes likely to be found? What type of volcano is found in each area, and why?

4. Why are earthquakes and volcanoes often found in the same areas?

Use What You've Learned

5. Many scientists believe that the Great Rift Valley of Africa will be the location of the next ocean as eastern Africa breaks away from the rest of the continent. Research what is known and predicted about the movement of plates in this area.

www·science·nelson·com

6. Why might frequent, smaller earthquakes be desirable along a transform fault boundary?

7. If a small earthquake occurs in Ontario, should people be worried that it is a sign of a bigger earthquake to come? Is this a reasonable worry?

8. What is a seismograph? What do scientists use seismographs for? Research seismographs and their use. Can you make a simple seismograph from everyday materials?

www·science·nelson·com

9. In Iceland, where hot magma is close to Earth's surface, many people use its heat to warm their homes. This form of energy is called geothermal energy. Research geothermal energy. Would geothermal energy be practical to use in your community? Prepare a report.

www·science·nelson·com

10. Research and prepare a report on a volcanic eruption. You might consider Thera in 2600 B.C.E., Mount Vesuvius in 79 C.E., Krakatoa in 1883, Mount Pelee in 1902, Mount St. Helens in 1980 or 2004, or the Nisga'a lava beds. Is there a volcano erupting now somewhere in the world that you could research?

www·science·nelson·com

Think Critically

11. Imagine that you had to live and work in one of the following areas: an area where earthquakes are common or an area where volcanic eruptions are common. Which area would you choose? Give reasons for your choice.

12. Describe some of the social problems that could result if scientists predicted an earthquake near a city. What could be done in advance to minimize these problems?

Reflect on Your Learning

13. Which hands-on activity in this chapter was most helpful to your learning? How did it help you learn?

14. How have your ideas about Earth changed since learning about plate movements? Will you look at the world differently since studying plate tectonics?

Making Connections

Writing the Life Story of a Rock

Looking Back

In this unit, you studied examples of both sudden and gradual changes to Earth's crust. You learned about rock and how rock from one family can change into rock from another family. You also learned that Earth's crust is made up of moving plates.

Throughout the unit, you explored, created, and used models to help you learn about processes that were too large or dangerous, or happened too slowly, for classroom observations.

In this activity, you will work with a partner to develop a detailed presentation about a single rock.

Demonstrate Your Learning

Part 1: Investigate the present.

Select a rock from the samples provided by your teacher. If you are interested in using a rock you found, check with your teacher. Use physical properties to identify your rock. Make notes for your presentation as you work. Consider the following questions about your rock.

- What rock family does your rock belong to?
- Where is it usually found?
- Are there ways in which people use this type of rock?
- Is there anything especially interesting or unusual about this type of rock?

You might do Internet research to help you find out more about your rock.

www·science·nelson·com **GO**

Part 2: Describe a realistic past

Take notes as you discuss your ideas about a realistic past for your rock. Consider the following questions when describing your rock's past.

- How and where do you think your rock was formed?
- Could it once have been a different type of rock?
- How could it have been changed since it was formed?
- How could it have been moved from where it was formed?
- What type of land feature could it have been part of?

Remember that your rock's past may cover several million years. Create a model to explain your rock's past to your classmates.

Part 3: Describe a realistic future

Take notes as you brainstorm the possibilities for the future of your rock. Consider the following questions when describing your rock's future.

- What realistic predictions could you make about what will happen to your rock over the next several million years?
- Could it travel far from where it is now?
- Could it be dramatically changed?

Create a model to demonstrate your rock's possible future.

Part 4: Develop your presentation

Use your notes to develop a presentation. Your presentation could be oral or written. It could be a book, a Web site, a skit, a comic strip, or anything else you choose. Use your creativity to develop a presentation that will be informative and interesting for your classmates. Use simple materials to make models that will help you explain some of the processes you describe.

> **▶ ASSESSMENT** ⊗
>
> **Check to make sure that your work provides evidence that you are able to**
>
> - identify and classify your rock
> - describe how your rock was formed
> - describe the possible effects of weathering, erosion, and plate movement on your rock
> - create models to help you explain processes that could have, or might in the future, affect your rock
> - use appropriate scientific language to describe these processes

SKILLS HANDBOOK

THINKING AS A SCIENTIST

WORKING AS A SCIENTIST

READING FOR INFORMATION

COMMUNICATING IN SCIENCE

THINKING AS A SCIENTIST

You may not think you're a scientist, but you are! You investigate the world around you, just like scientists do. When you investigate, you are looking for answers. Imagine that you are planning to buy a mountain bike. You want to find out which model is the best buy. First, you write a list of questions. Then you visit stores, check print and Internet sources, and talk to your friends to find the answers. You are conducting an investigation.

Scientists conduct investigations for different purposes:

- *Scientists investigate the natural world in order to describe it.* For example, scientists study rocks to find out what their properties are, how they were formed, and how they are still changing today.

- *Scientists investigate how objects and organisms can be classified.* For example, scientists examine substances and classify them as pure substances or mixtures.

- *Scientists investigate to test their ideas about the natural world.* Scientists ask cause-and-effect questions about what they observe. They propose hypotheses to answer their questions. Then they design experiments to test their hypotheses.

CONDUCTING AN INVESTIGATION

When you conduct an investigation or design an experiment, you will use a variety of skills. Refer to this section when you have questions about how to use any of the following investigation skills and processes.

- Questioning
- Predicting
- Hypothesizing
- Controlling Variables
- Observing
- Measuring
- Classifying
- Inferring
- Interpreting Data
- Communicating
- Creating Models

Questioning

Scientific investigations start with good questions. To write a good question, you must first decide what you want to know. This will help you think of, or formulate, a question that will lead you to the information you want.

You must think carefully about what you want to know in order to develop a good question. The question should include the information you want to find out.

Sometimes an investigation starts with a special type of question, called a cause-and-effect question. A cause-and-effect question asks whether something is causing something else. It might start in one of the following ways:

What causes ...?
How does ... affect ...?
What would happen if ...?

When an investigation starts with a cause-and-effect question, it also has a hypothesis. Read "Hypothesizing" on page 270 to find out more about hypotheses.

PRACTICE

Think of some everyday examples of cause and effect, and write statements about them. Here's one example: "When I stay up too late, I'm tired the next day." Then turn your statements into cause-and-effect questions: for example, "What would happen if I stayed up late?"

Predicting

A prediction states what is likely to happen based on what is already known. Scientists base their predictions on their observations. They look for patterns in the data they gather to help them see what might happen next or in a similar situation. This is how meteorologists come up with weather forecasts.

Remember that predictions are not guesses. They are based on solid evidence and careful observations. You must be able to give reasons for your predictions. You must also be able to test them by doing experiments.

Hypothesizing

Figure 1
This student is conducting an investigation to test this hypothesis: if the number of times the balloon is rubbed against hair increases, then the length of time it will stick to the wall will increase.

To test your questions and predictions scientifically, you need to conduct an investigation. Use a question or prediction to create a cause-and-effect statement that can be tested. This kind of statement is called a **hypothesis**.

An easy way to make sure that your hypothesis is a cause-and-effect statement is to use the form "If … then …." If the independent variable (cause) is changed, then the dependent variable (effect) will change in a specific way (**Figure 1**). For example, "If the number of times a balloon is rubbed against hair (the cause or independent variable) is increased, then the length of time it sticks to a wall (the effect or dependent variable) increases." Read "Controlling Variables" below to find out more about independent and dependent variables.

Questions, predictions, and hypotheses go hand in hand. For example, your question might be "Does a balloon stick better if you rub it more times on your hair?" Your prediction might be "A balloon will stick to a wall longer the more times it is rubbed on your hair." Your hypothesis might be "If the number of times you rub a balloon on your hair is increased, then the length of time it sticks to a wall will increase." If you prove that your hypothesis is correct, then you have confirmed your prediction.

You can create more than one hypothesis from the same question or prediction. Another student might test the hypothesis "If the number of times you rub a balloon on your hair is increased, then the length of time it sticks to a wall will be unchanged."

Of course, both of you cannot be correct. When you conduct an investigation, you do not always prove that your hypothesis is correct. Sometimes you prove that your hypothesis is incorrect. An investigation that proves your hypothesis to be incorrect is not a bad investigation or a waste of time. It has contributed to your scientific knowledge. You can re-evaluate your hypothesis and design a new experiment.

PRACTICE

Write hypotheses for questions or predictions about rubbing a balloon on your hair and sticking it to a wall. Start with the questions above, and then write your own questions. For example, if your question is "Does the balloon stick better if you rub it more times?", then your hypothesis might be "If the number of times you rub the balloon on your hair is increased, then the length of time it sticks to the wall is increased."

Controlling Variables

When you are planning an investigation, you need to make sure that your results will be reliable by conducting a fair test. To make sure that an investigation is a fair test, scientists identify all the variables that might affect their results. Then they make sure that they change only one variable at a time. This way they know that their results are caused by the variable they changed and not by any other variables (**Figure 2**).

- The variable that is changed in an investigation is called the **independent variable**.
- The variable that is affected by a change is called the **dependent variable**. This is the variable you measure to see how it was affected by the independent variable.
- All the other conditions that remain unchanged in an experiment, so that you know they did not have any effect on the outcome, are called **controlled variables**.

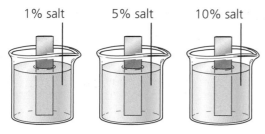
1% salt   5% salt   10% salt

Figure 2
This investigation was designed to find out if the amount of salt in a solution has an effect on the rusting of metal.
- The amount of salt in each solution is the independent variable.
- The amount of rust on the pieces of metal is the dependent variable.
- The amount of water in each beaker and the amount of time the metal strip stays in the water are two of the controlled variables.

PRACTICE

Suppose that you have noticed mould growing on an orange. You want to know what is causing the mould. What variables will you have to consider in order to design a fair test? Which variable will you try changing in your test? What is this variable called? What will your dependent variable be? What will your controlled variables be?

Observing

When you observe something, you use your senses to learn about the world around you. You can also use tools, such as a balance, metre stick, and microscope.

Some observations are measurable. They can be expressed in numbers. Observations of time, temperature, volume, and distance can all be measured. These types of observations are called **quantitative observations**.

Other observations describe qualities that cannot be measured. The smell of a fungus, the shape of a flower petal, or the texture of soil are all examples of qualities that cannot be put in numbers. These types of observations are called **qualitative observations**. Qualitative observations also include colour, taste, clarity, and state of matter.

The colour and shape of this box are qualitative observations. The measurements of its height, depth, and width are quantitative observations.

PRACTICE

Make a table with two columns, one for quantitative observations and the other for qualitative observations. Find a rock that you think is interesting. See if you can make 10 observations about the rock. Record your observations in your table.

Measuring

Measuring is an important part of observation. When you measure an object, you can describe it precisely and keep track of any changes. To learn about using measuring tools, turn to "Measurement and Measuring Tools" on page 285.

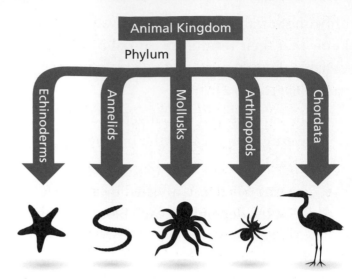

Figure 3

To help classify animals, scientists divide the animal kingdom into five smaller groups called *phyla* (singular *phylum*).

Measuring accurately requires care.

Classifying

You classify things when you sort them into groups based on their similarities and differences. When you sort clothes, sporting equipment, or books, you are using a classification system. To be helpful to other people, a classification system must make sense to them. If, for example, your local supermarket sorted all the products in alphabetical order, so that soap, soup, and soy sauce were all on the same shelf, no one would be able to find anything!

Classification is an important skill in science. Scientists try to group objects, organisms, and events in order to understand the nature of life (**Figure 3**).

PRACTICE

Gather photos of 15 to 20 different insects, seashells, or flowers. Try to include as much variety as possible. How are all your samples alike? How are they different? How could you classify them?

Inferring

An inference is a possible explanation of something you observe. It is an educated guess based on your experience, knowledge, and observations. You can test your inferences by doing experiments.

It is important to remember that an inference is only an educated guess. There is always some uncertainty. For example, if you hear a dog barking but do not see the dog, you may infer that it is your neighbour's dog. It may, however, be some other dog that sounds the same. An observation, on the other hand, is based on what you discover with your senses and measuring tools. If you say that you heard a dog barking, you are making an observation.

Decide whether each of these statements is an observation or an inference.

a) You see a bottle filled with clear liquid. You conclude that the liquid is water.

b) You notice that your head is stuffed up and you feel hot. You decide that you must have a cold.

c) You tell a friend that three new houses are being built in your neighbourhood.

d) You see a wasp crawling on the ground instead of flying. You conclude that it must be sick.

e) You notice that you are thirsty after playing sports.

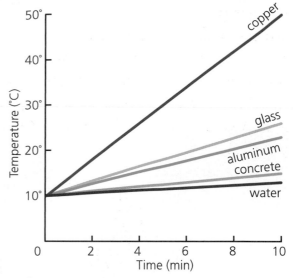

Figure 4

This graph shows data from an investigation about the heating rates of different materials. What patterns and relationships can you see from this data?

Interpreting Data

When you interpret data from an investigation, you make sense of it. You examine and compare the measurements you have made. You look for patterns and relationships that will help you explain your results and give you new information about the question you are investigating. Once you have interpreted your data, you can tell whether your predictions or hypothesis are correct. You may even come up with a new hypothesis that can be tested in a new experiment.

Often, making tables or graphs of your data will help you see patterns and relationships more easily (**Figure 4**). Turn to "Communicating in Science" on page 297 to learn more about creating data tables and graphing your results.

Communicating

Scientists learn from one another by sharing their observations and conclusions. They present their data in charts, tables, or graphs and in written reports. In this student text, each investigation or activity tells you how to prepare and present your results. To learn more about communicating in a written report, turn to "Writing a Lab Report" on page 300.

Creating Models

Have you ever seen a model of the solar system? Many teachers use a small model of the solar system when teaching about space because it shows how the nine planets orbit the Sun. The concept of how planets orbit the Sun is very difficult to imagine without being able to see it.

A scientific model is an idea, illustration, or object that represents something in the natural world (**Figure 5** on page 274). Models allow you to examine and investigate things that are very large, very small, very complicated, very dangerous, or hidden from view. They also allow you to investigate processes that happen too slowly to be observed directly. You can model, in a few minutes, processes that take months or even millions of years to occur.

Figure 5

Why do we use these models? How are they different from what they represent? Are there any limitations or disadvantages to using them? Think of another model you could make to represent each of these things.

A model of the solar system is an example of a physical model. You can create physical models from very simple materials. Have you ever thrown a paper airplane? If so, you were actually testing a model of a real airplane. You could use paper airplane models to test different airplane designs.

Illustrations are also models. A map of Earth, showing all the biomes, is a model. So is a drawing of a particle of water. Models can be created from ideas and words, as well. Some Aboriginal stories communicate models of interconnected ecosystems and the appropriate place of humans in nature. The particle model explains, in words, what matter

is made from and why different substances behave as they do.

Although models have many advantages, they also have some disadvantages. They are usually more simple than what they represent.

Models change over time as scientists make new observations. For example, models of Earth have changed. Long ago, European people thought that Earth was flat. They thought that if they sailed far enough out to sea, they would fall off the edge. Central American people thought that Earth was held up by a turtle. When the turtle moved, Earth rumbled. As scientists made more and more observations over time, they revised their model of Earth.

SOLVING A PROBLEM

Refer to this section when you are doing a "Solve a Problem" activity.

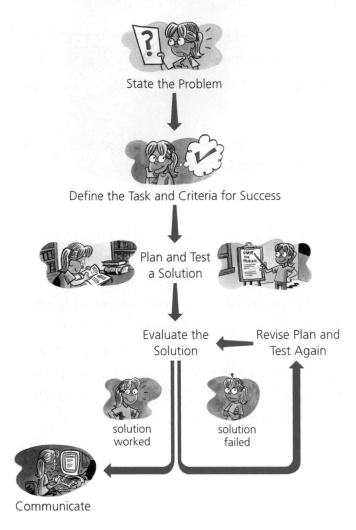

State the Problem

Define the Task and Criteria for Success

Plan and Test a Solution

Evaluate the Solution ← Revise Plan and Test Again

solution worked   solution failed

Communicate

State the Problem

The first step in solving a problem is to state what the problem is. Imagine, for example, that you are part of a group that is investigating how to reduce the risk of people getting the West Nile Virus. People can become very sick from this virus.

When you are trying to understand a problem, ask yourself these questions:

- What is the problem? How can I state it as a problem?

- What do I already know about the problem?
- What do I need to know to solve the problem?

Define the Task and the Criteria for Success

Once you understand the problem, you can define the task. The task is what you need to do to find a solution. For the West Nile Virus problem, you may need to find a way to reduce the number of mosquitoes in your community because they could be carrying the West Nile Virus.

Before you start to consider possible solutions, you need to know what you want your solution to achieve. One of the criteria for success is fewer mosquitoes. Not every solution that would help you achieve success will be acceptable, however. For example, some chemical solutions may kill other, valuable insects or may be poisonous to birds and pets. The solution should not be worse than the problem it is meant to solve. As well, there are limits on your choices. These limits may include the cost of the solution, the availability of materials, and safety.

Use the following questions to help you define your task and your criteria for success:

- What do I want my solution to achieve?
- What criteria should my solution meet?
- What are the limits on my solution?

Plan and Test a Solution

The planning stage is when you look at possible solutions and decide which solution is most likely to work. This stage usually starts with brainstorming possible solutions. When you are looking for solutions, let your imagination go. Keep a record of your ideas. Include sketches, word webs, and other graphic organizers to help you.

As you examine the possible solutions, you may find new questions that need to be researched. You may want to do library and Internet research, interview experts, and talk to people in your community about the problem.

Choose one solution to try. For the West Nile Virus problem, you may decide to inspect your community for wet areas where mosquitoes breed, and try to eliminate as many of these wet areas as possible. You have discovered, through your research, that this solution is highly effective for reducing mosquito populations. It also has the advantage of not involving chemicals and costing very little.

Now make a list of the materials and equipment you will need. Develop your plan on paper so that other people can examine it and add suggestions. Make your plan as thorough as possible so that you have a blueprint for how you are going to carry out your solution. Show your plan to your teacher for approval.

Once your teacher has approved your plan, you need to test it. Testing allows you to see how well your plan works and to decide whether it meets your criteria for success. Testing also tells you what you might need to do to improve your solution.

Evaluate the Solution

The evaluating stage is when you consider how well your solution worked. Use these questions to help you evaluate your solution:

- What worked well? What did not work well?
- What would I do differently next time?
- What did I learn that I can apply to other problems?

If your solution did not work, go back to your plan and revise it. Then test again.

Communicate

At the end of your problem-solving activity, you should have a recommendation to share with others. To communicate your recommendation, you need to write a report. Think about what information you should include in your report. For example, you may want to include visuals, such as diagrams and tables, to help others understand your results and recommendation.

DESIGNING YOUR OWN EXPERIMENT

Refer to this section when you are designing your own experiment.

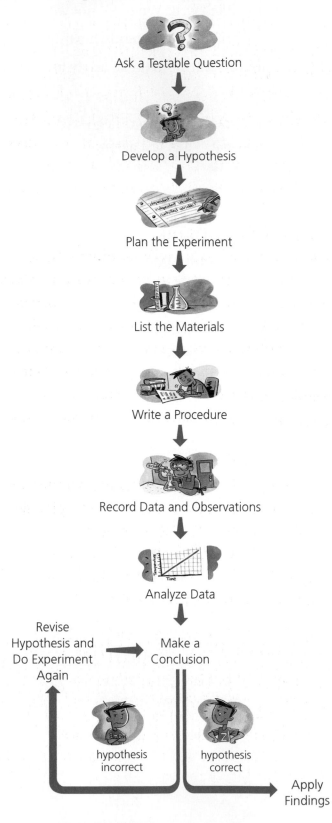

Ask a Testable Question

Develop a Hypothesis

Plan the Experiment

List the Materials

Write a Procedure

Record Data and Observations

Analyze Data

Revise Hypothesis and Do Experiment Again → Make a Conclusion

hypothesis incorrect

hypothesis correct

Apply Findings

After observing the difference between his lunch and Dal's, Simon wondered why his food was not as fresh as Dal's.

Scientists design experiments to test their ideas about the things they observe. They follow the same steps you will follow when you design an experiment.

Ask a Testable Question

The first thing you need is a testable question. A testable question is a question that you can answer by conducting a test. A good, precise question will help you design your experiment. What question do you think Simon, in the picture above, would ask?

A testable question is often a cause-and-effect question. Turn to "Questioning" on page 269 to learn how to formulate a cause-and-effect question.

Develop a Hypothesis

Next, use your past experiences and observations to formulate a hypothesis. Your hypothesis should provide an answer to your question and briefly explain why you think the answer is correct. It should be testable through an experiment. What do you think Simon's hypothesis would be? Turn to "Hypothesizing" on page 270 to learn how to formulate a hypothesis.

Plan the Experiment

Now you need to plan how you will conduct your experiment. Remember that your experiment must be a fair test. Also remember that you must only change one independent variable at a time. You need to know what your dependent variable will be and what variables you will control. What do you think Simon's independent variable would be? What do you think his dependent variable would be? What variables would he need to control? Turn to "Controlling Variables" on page 270 to learn about fair tests and variables.

List the Materials

Make a list of all the materials you will need to conduct your experiment. Your list must include specific quantities and sizes, where needed. As well, you should draw a diagram to show how you will set up the equipment. What materials would Simon need to complete his experiment?

Write a Procedure

The procedure is a step-by-step description of how you will perform your experiment. It must be clear enough for someone else to follow exactly. It must explain how you will deal with each of the variables in your experiment. As well, it must include any safety precautions. Your teacher must approve your procedure and list of materials. What steps and safety precautions should Simon include?

Record Data and Observations

You need to make careful observations, so that you can be sure about the effects of the independent variable. Record your observations, both qualitative and quantitative, in a data table, tally chart, or graph. How would Simon record his observations?

Turn to "Observing" on page 271 to read about qualitative and quantitative observations. Turn to "Creating Data Tables" on page 297 to read about creating data tables.

Analyze Data

If your experiment is a fair test, you can use your observations to determine the effects of the independent variable. You can analyze your observations to find out how the independent and dependent variables are related. Scientists often conduct the same test several times to make sure that their observations are accurate.

Make a Conclusion

When you have analyzed your observations, you can use the results to answer your question and determine if your hypothesis was correct. You can feel confident about your conclusion if your experiment was a fair test and there was little room for error. If you proved that your hypothesis was incorrect, you can revise your hypothesis and perform the experiment again.

Apply Findings

The results of scientific experiments add to our knowledge about the world. For example, the results may be applied to develop new technologies and medicines, which help to improve our lives. How do you think Simon could use what he discovered?

PRACTICE

You are a tennis player. You observe that your tennis ball bounces differently when the court is wet. Design a fair test to investigate your observation. Use the headings in this section.

EXPLORING AN ISSUE

Use this section when you are doing an "Explore an Issue" activity.

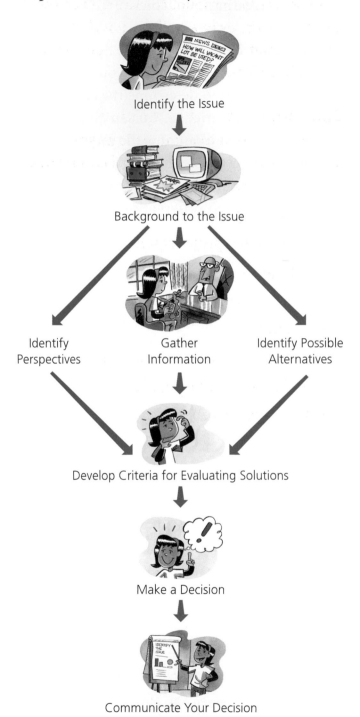

Identify the Issue

Background to the Issue

Identify Perspectives

Gather Information

Identify Possible Alternatives

Develop Criteria for Evaluating Solutions

Make a Decision

Communicate Your Decision

An issue is a situation in which several points of view need to be considered in order to make a decision. Often what different people think is the best decision is based on what they think is important or on what they value. Often, it is difficult to come to a decision that everyone agrees with.

When a decision has an impact on many people or on the environment, it is important to explore the issue carefully. This means thinking about all the possible solutions and trying to understand all the different points of view—not just your own point of view. It also means researching and investigating your ideas, and talking to and listening to others.

Identify the Issue

The first step in exploring an issue is to identify what the issue is. An issue has more than one solution, and there are different points of view about which solution is the best. Try stating the issue as a question: "What should ...?

Background to the Issue

The background to the issue is all the information that needs to be gathered and considered before a decision can be made.

- *Identify perspectives.* There are always different points of view on an issue. That's what makes it an issue. For example, suppose that your municipal council is trying to decide how to use some vacant land next to your school. You and other students have asked the council to zone the land as a nature park. Another group is proposing that the land be used to build a seniors' home because there is a shortage of this kind of housing. Some school administrators would like to use the land to build a track for runners and sporting events.

- *Gather information.* The decision you reach must be based on a good understanding of the issue. You must be in a position to choose the most appropriate solution. To do this, you need to gather factual information that represents all the different points of view. Watch out for biased information, presenting only one side of the issue. Develop good questions and a plan for your research. Your research may include talking to people, reading about the issue, and doing Internet research. For the land-use issue, you may also want to visit the site to make observations.

- *Identify possible alternatives.* After identifying points of view and gathering information, you can now generate a list of possible solutions. You might, for example, come up with the following solutions for the land-use issue:
  - Turn the land into a nature park for the community and the school.
  - Use the land as a playing field and track for the community and the school.
  - Create a combination park and playing field.
  - Use the land to build a seniors' home, with a "nature" garden.

Develop Criteria for Evaluating Solutions

Develop criteria to evaluate each possible solution. For example, should the solution be the one that has the most community support? Should it be the one that protects the environment? You need to decide which criteria you will use to evaluate the solutions so that you can decide which solution is the best.

Make a Decision

This is the stage where everyone gets a chance to share his or her ideas and the information he or she gathered about the issue. Then the group needs to evaluate all the possible solutions and decide on one solution based on the list of criteria.

Communicate Your Decision

Choose a method to communicate your decision. For example, you could choose one of the following methods:

- Write a report.
- Give an oral presentation.
- Design a poster.
- Prepare a slide show.
- Create a video.
- Organize a panel presentation.
- Write a newspaper article.
- Hold a formal debate.

WORKING AS A SCIENTIST

GETTING OFF TO A SAFE START

Science activities and investigations can be a lot of fun. You have the chance to work with new equipment and substances. These can be dangerous, however, so you have to pay attention! You also have to know and follow special rules. Here are the most important rules to remember.

1 Follow your teacher's directions.

- Listen to your teacher's directions, and follow them carefully.

- Ask your teacher for directions if you are not sure what to do.

- Never change anything, or start an activity on your own, without your teacher's approval.

- Get your teacher's approval before you start an experiment that you have designed yourself.

2 Act responsibly.

- Pay attention to your own safety and the safety of others.

- Tell your teacher immediately if you see a safety hazard, such as broken glass or a spill. Also tell your teacher if you see another student doing something that you think is dangerous.

- Tell your teacher about any allergies or medical problems you have, or about anything else your teacher should know.

- Do not wear contact lenses while doing experiments.

- Read all written instructions carefully before you start an activity.

- Clean up and put away any equipment after you are finished.

3 Be science-ready.

- Come prepared with your student text, notebook, pencil, worksheets, and anything else you need for an activity or investigation.

- Keep yourself and your work area tidy and clean.

- Wash your hands carefully with soap and water at the end of each activity or investigation.

- Never eat, drink, or chew gum in the science classroom.

- Wear safety goggles or other safety equipment when instructed by your teacher.

- Keep your clothing and hair out of the way. Roll up your sleeves, tuck in loose clothing, and tie back loose hair. Remove any loose jewellery.

SAFE SCIENCE

Follow these instructions to use chemicals and equipment safely in the science classroom.

HEAT, FIRE, AND ELECTRICITY

- Never heat anything without your teacher's permission.
- Always wear safety goggles when you are working with fire.
- Keep yourself, and anything else that can burn, away from heat or flames.
- Never reach across a flame.
- Before you heat a test tube or another container, point it away from yourself and others. Liquid inside can splash or boil over when heated.
- Never heat a liquid in a closed container.
- Use tongs or heat-resistant gloves to pick up a hot object.
- Test an object that has been heated before you touch it. Slowly bring the back of your hand toward the object to make sure that it is not hot.
- Know where the fire extinguisher and fire blanket are kept in your classroom.
- Never touch an electrical appliance or outlet with wet hands.
- Keep water away from electrical equipment.

CHEMICALS

- If you spill a chemical (or anything else), tell your teacher immediately.
- Never taste, smell, touch, or mix chemicals without your teacher's permission.
- Never put your nose directly over a chemical to smell it. Gently wave your hand over the chemical until you can smell the fumes.
- Keep the lids on chemicals you are not using tightly closed.
- Wash your hands well with soap after handling chemicals.
- Never pour anything into a sink without your teacher's permission.
- If any part of your body comes in contact with a chemical, wash the area immediately and thoroughly with water. If your eyes are affected, do not touch them but wash them immediately and continuously with cool water for at least 15 min. Inform your teacher.

HANDLE WITH CARE

GLASS AND SHARP OBJECTS

- Handle glassware, knives, and other sharp instruments with extra care.
- If you break glassware or cut yourself, tell your teacher immediately.
- Never work with cracked or chipped glassware. Give it to your teacher.
- Use knives and other cutting instruments carefully. Never point a knife or sharp object at another person.
- When cutting, make sure that you cut away from yourself and others.

LIVING THINGS

- Treat all living things with care and respect.
- Never treat an animal in a way that would cause it pain or injury.
- Touch animals only when necessary. Follow your teacher's directions.
- Always wash your hands with soap after working with animals or touching their cages or containers.

Caution Symbols

The activities and investigations in *B.C. Science Probe 7* are safe to perform, but accidents can happen. This is why potential safety hazards are identified with caution symbols and red type (**Figure 1**). Make sure you read the cautions carefully and understand what they mean. Check with your teacher if you are unsure.

Safety Symbols

The following safety symbols are used throughout Canada to identify products that can be hazardous (**Figures 2** and **3**). Make sure that you know what each symbol means. Always use extra care when you see any of these symbols in your classroom or anywhere else.

 Wash your hands with soap and water after each time you work with the plants.

Figure 1
Potential safety hazards are identified with caution symbols and red type.

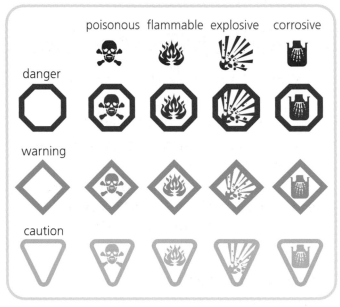

Figure 2
Hazardous Household Product Symbols (HHPS) appear on many products that are used in the home. Different shapes show the level of danger.

Figure 3
Workplace Hazardous Materials Information System (WHMIS) symbols identify dangerous materials that are used in all workplaces, including schools.

PRACTICE

In a group, create a safety poster for your classroom. For example, you could create a map of the route your class should follow when a fire alarm sounds, a map of where safety materials (such as a fire extinguisher and a first-aid kit) are located in your classroom, information about the safe use of a specific tool, or a list of safety rules.

MEASUREMENT AND MEASURING TOOLS

Refer to this section when you need help with taking measurements.

Measuring is an important part of doing science. Measurements allow you to give exact information when you are describing something.

These are the most commonly used measurements:

- Length
- Mass
- Volume
- Temperature

The science community and most countries in the world, including Canada, use the SI system. The SI system is commonly called the metric system.

The metric system is based on multiples of 10. Larger and smaller units are created by multiplying or dividing the value of the base units by multiples of 10. For example, the prefix *kilo-* means "multiplied by 1000." Therefore, one kilometre is equal to one thousand metres. The prefix *milli-* means "divided by 1000," so one millimetre is equal to 1/1000 of a metre. Some common SI prefixes are listed in **Table 1**.

Table 1   Common SI Prefixes

Prefix	Symbol	Factor by which unit is multiplied	Example
kilo	k	1000	1 km = 1000 m
hecto	h	100	1 hm = 100 m
deca	da	10	1 dam = 10 m
		1	
deci	d	0.1	1 dm = 0.1 m
centi	c	0.01	1 cm = 0.01 m
milli	m	0.001	1 mm = 0.001 m

To convert from one unit to another, you simply multiply by a conversion factor. For example, to convert 12.4 m (metres) to centimetres (cm), you use the relationship 1 cm = 0.01 m, or $1 \text{ cm} = \frac{1}{100}$ m.

$$12.4 \text{ m} = ? \text{ cm}$$
$$1 \text{ cm} = 0.01 \text{ m}$$
$$(12.4 \text{ m}) \left(\frac{1 \text{ cm}}{0.01 \text{ m}} \right) = 1240 \text{ cm}$$

Any conversion between quantities with the same base unit can be done like this, once you know the conversion factor.

PRACTICE

a) Convert 23 km (kilometres) to metres (m) and to millimetres (mm).

b) Convert 675 mL (millilitres) to litres (L).

c) Convert 450 g (grams) to kilograms (kg) and to milligrams (mg).

If you are not sure which conversion factor you need, look at the information in the box below and in the boxes on pages 286 and 287.

Measuring Length

Length is the distance between two points. Four units can be used to measure length: kilometres (km), metres (m), centimetres (cm), and millimetres (mm).

| 10 mm = 1 cm | 100 cm = 1 m |
| 1000 mm = 1 m | 1000 m = 1 km |

You measure length when you want to find out how long something is. You also measure length when you want to know how deep, how tall, how far, or how wide something is. The metre is the basic unit of length (**Figure 4** on the next page).

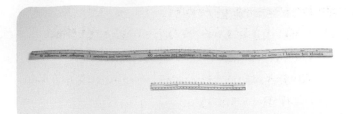

Figure 4
Metric rulers are used to measure lengths in millimetres and centimetres, up to 30 cm. Metre sticks measure longer lengths, up to 100 cm.

PRACTICE

Which unit—millimetres, centimetres, metres, or kilometres—would you use to measure each quantity?

a) the width of a scar or mole on your body

b) the length that your toenails grow in one month

c) your height

d) the length that your hair grows in one month

e) the distance between your home and Calgary

f) the distance between two planets

Tips for Measuring Length

- Always start measuring from the zero mark on a ruler, not from the edge of the ruler.

- Look directly at the lines on the ruler. If you try to read the ruler at an angle, you will get an incorrect measurement.

- To measure something that is not in a straight line, use a piece of string (**Figure 5**). Cut or mark the string. Then use a ruler to measure the length of the string. You could also use a tape measure made from fabric.

Figure 5

Measuring Volume

Volume is the amount of space that something takes up. The volume of a solid is usually measured in cubic metres (m³) or cubic centimetres (cm³). The volume of a liquid is usually measured in litres (L) or millilitres (mL).

1000 mL = 1 L	1 L = 1000 cm³
1 cm³ = 1 mL	1000 L = 1 m³

The volume of a rectangular solid is calculated by measuring the length, width, and height of the solid and then by using the formula

Volume = length × width × height

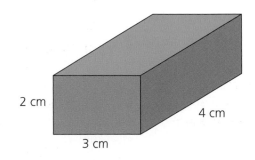

2 cm

3 cm

4 cm

Volume is also used to measure the amount of liquid in a container. Scientists use special containers, such as beakers and graduated cylinders, to get precise measurements of volume.

You can also use liquid to help measure the volume of irregularly shaped solids, such as rocks. To measure the volume of an irregularly shaped solid, choose a container (such as a graduated cylinder) that the irregular solid will fit inside. Pour water into the empty container until it is about half full. Record the volume of water in the container, and then carefully add the solid. Make sure that the solid is completely submerged in the water. Record the volume of the water plus the solid. Calculate the volume of the solid using the following formula:

Volume of solid = (volume of water + solid) − volume of water

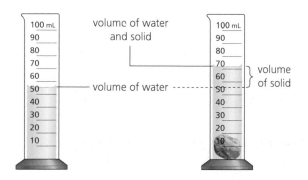

PRACTICE

What volume of liquids do you drink in an average day? Use the illustrations of volume measurements to help you answer this question.

measuring cup   milk carton   tablespoon   pop bottle
   500 mL   1 L   15 mL   2 L

Tips for Measuring Volume

- Use a beaker that is big enough to hold twice as much liquid as you need. You want a lot of space so that you can get an accurate reading.

- To measure liquid in a graduated cylinder (or a beaker or a measuring cup), make sure that your eyes are at the same level as the top of the liquid. You will see that the surface of the liquid curves downward. This downward curve is called the **meniscus**. You need to measure the volume from the bottom of the meniscus (**Figure 6**).

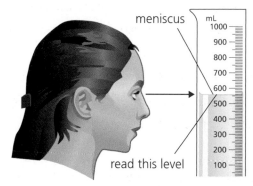

Figure 6
Reading the measurement of a liquid correctly

- Use a graduated cylinder to get the most accurate measurement of volume.

Measuring Mass

Mass is the amount of matter in an object. In everyday life, weight is often confused with mass. For example, you probably state your weight in kilograms. In fact, what you are really stating is your mass. The units that are used to measure mass are grams (g), milligrams (mg), kilograms (kg), and metric tonnes (t).

1000 g = 1 kg   1000 kg = 1 t
1000 mg = 1 g

Skills Handbook

Scientists use balances to measure mass. Two types of balances are the triple-beam balance (**Figure 7**) and the platform, or equal-arm, balance (**Figure 8**).

Figure 7

A triple-beam balance: Place the object you are measuring on the pan. Adjust the weights on each beam (starting with the largest) until the pointer on the right side is level with the zero mark. Then add the values of each beam to find the measurement.

Figure 8

A platform balance: Place the object you are measuring on one pan. Add weights to the other pan until the two pans are level. Then add the values of the weights you added. The total will be equal to the mass of the object you are measuring.

Tips for Measuring Mass
- To measure the mass of a liquid, first measure the mass of a suitable container. Then measure the mass of the liquid in the container. Subtract the mass of the container from the mass of the liquid and the container.

- To measure the mass of a powder or crystals, first determine the mass of a sheet of paper. Then place the sample on the sheet of paper, and measure the mass of both. Subtract the mass of the paper from the mass of the sample and the sheet of paper.

Measuring Temperature

Temperature is the degree of hotness or coldness of an object. In science, temperature is measured in degrees Celsius.

0°C = freezing point of water
20°C = warm spring day
37.6°C = normal body temperature
65°C = water hot to touch
100°C = boiling point of water

Measuring the temperature of water

Each mark on a Celsius thermometer is equal to one degree Celsius. The glass contains a coloured liquid—usually mercury or alcohol. When you place the thermometer in a substance, the liquid in the thermometer moves to indicate the temperature.

Tips for Measuring Temperature
- Make sure that the coloured liquid has stopped moving before you take your reading.

- Hold the thermometer at eye level to be sure that your reading is accurate.

READING FOR INFORMATION

USING GRAPHIC ORGANIZERS

Diagrams that are used to organize and display ideas visually are
called graphic organizers. A graphic organizer can help you see
connections and patterns among different ideas. Different graphic
organizers are used for different purposes.

- To Show Processes
- To Organize Ideas and Thinking
- To Compare and Contrast
- To Show Properties or Characteristics

To Show Processes

You can use a **flow chart** to show
a sequence of steps or a time line.

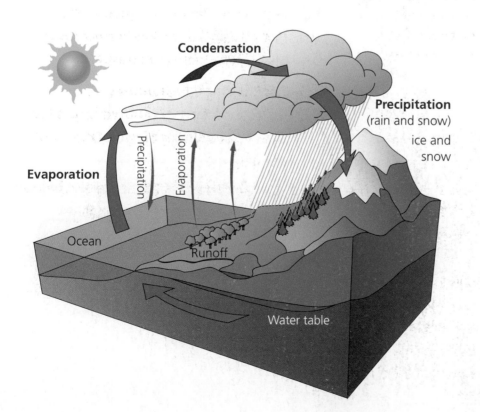

You can use a **cycle map** to show
cycles in nature.

To Organize Ideas and Thinking

A **concept map** is a collection of words or pictures, or both, connected with lines or arrows. You can write on the lines or arrows to explain the connections. You can use a concept map to brainstorm what you already know, to mind map your thinking, or to summarize what you have learned.

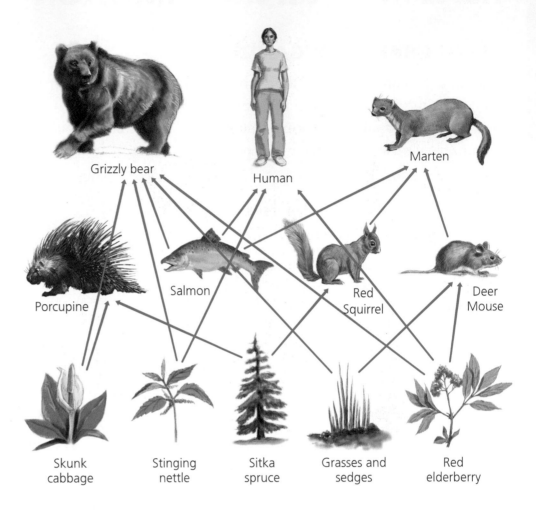

Grizzly bear

Human

Marten

Porcupine

Salmon

Red Squirrel

Deer Mouse

Skunk cabbage

Stinging nettle

Sitka spruce

Grasses and sedges

Red elderberry

You can use a **tree diagram** to show concepts that can be broken down into smaller categories.

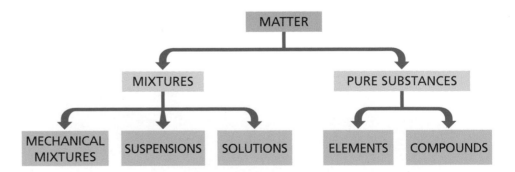

MATTER

MIXTURES

PURE SUBSTANCES

MECHANICAL MIXTURES

SUSPENSIONS

SOLUTIONS

ELEMENTS

COMPOUNDS

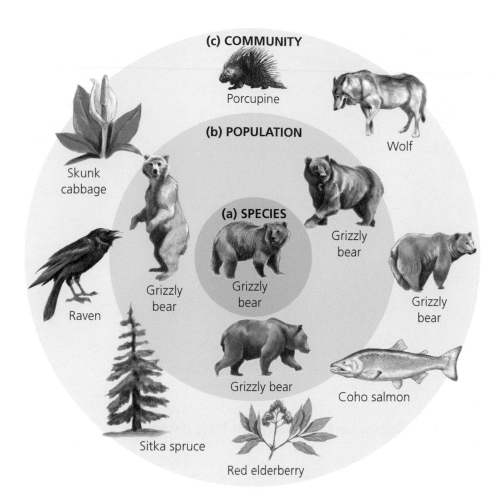

(c) COMMUNITY

Porcupine

Wolf

(b) POPULATION

Skunk cabbage

(a) SPECIES

Grizzly bear

Grizzly bear

Grizzly bear

Grizzly bear

Grizzly bear

Grizzly bear

Raven

Sitka spruce

Red elderberry

Coho salmon

You can use a **nested circle diagram** to show parts within a whole.

To Compare and Contrast

You can use a **comparison matrix** to record and compare observations or results.

Comparison of the Three States of Matter

State	Fixed mass?	Fixed volume?	Fixed shape?
solid	X	X	X
liquid	X	X	
gas	X		

You can use a **Venn diagram** to show similarities and differences. Similarities go in the middle section.

Comparing Plant and Animal Cells

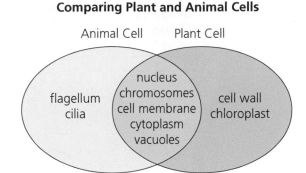

You can use a **compare and contrast chart** to show both similarities and differences.

Convergent and Divergent Plate Boundaries

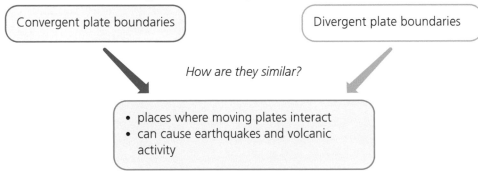

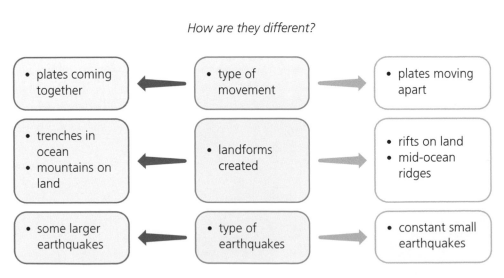

To Show Properties or Characteristics

You can use a **bubble map** to show properties.

READING STRATEGIES

The skills and strategies you use to help you read can differ, depending on the type of material you are reading. Reading a science text is different from reading a novel. When you are reading a science text, you are reading for information. Here are some strategies to help you read for information.

Before Reading

Skim the section you are going to read. Look at the illustrations, headings, and subheadings.

- *Preview.* What is this section about? How is it organized?

- *Make connections.* What do I already know about the topic? How is it connected to other topics I have already learned?

- *Predict.* What information will I find in this section? Which parts will give me the most information?

- *Set a purpose.* What questions do I have about the topic?

During Reading

Pause and think as you read. Spend time on the photographs, illustrations, tables, and graphs, as well as on the words.

- *Check your understanding.* What are the main ideas in this section? How would I explain them in my own words? What questions do I still have? Do I need to reread? Do I need to read more slowly, or can I read more quickly?

- *Determine the meanings of key science terms.* Can I figure out the meaning of unfamiliar terms from context clues in words or illustrations? Do I understand the definitions of terms in bold type? Is there something about the structure of a new term that will help me remember its meaning? Are there terms I should look up in the glossary?

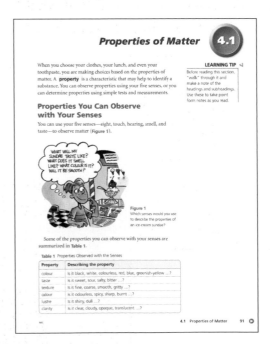

- *Make inferences.* What conclusions can I make from what I am reading? Can I make any conclusions by "reading between the lines"?

- *Visualize.* What mental pictures can I make to help me understand and remember what I am reading? Would it help to make a sketch?

- *Make connections.* How is this like things I already know?

- *Interpret visuals and graphics.* What additional information can I get from the photographs, illustrations, charts, or graphs?

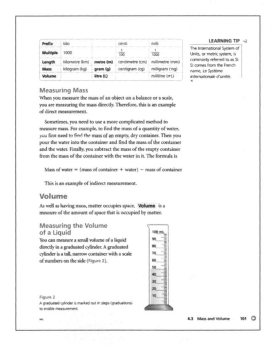

After Reading

Many of the strategies you use during reading can be used after reading as well. For example, in this text, there are questions to answer after you read. These questions will help you check your understanding and make connections.

- *Locate needed information.* Where can I find the information I need to answer the questions? Under what heading might I find the information? What terms in bold type should I skim for? What details do I need to include in my answers?

- *Synthesize.* How can I organize this information? What graphic organizer could I use? What headings or categories could I use?

- *React.* What are my opinions about this information? How does it, or might it, affect my life or my community? Do other students agree with my reactions?

- *Evaluate information.* What do I know now that I did not know before? Have any of my ideas changed as a result of what I have read? What questions do I still have?

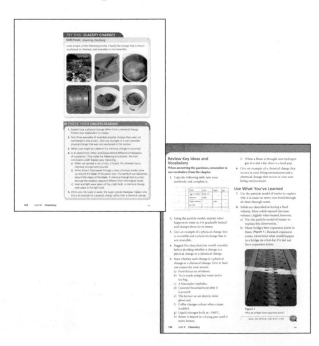

RESEARCHING

There is an incredible amount of scientific information that is available to you. Here are some tips on how to gather scientific information efficiently.

- Identify the Information You Need
- Find Sources of Information
- Evaluate the Sources of Information
- Record and Organize the Information
- Communicate the Information

Identify the Information You Need

Identify your research topic. Identify the purpose of your research.

Identify what you, or your group, already know about your topic. Also identify what you do not know. Develop a list of key questions that you need to answer. Identify categories based on your key questions. Use these categories to identify key search words.

Find Sources of Information

Identify all the places where you could look for information about your topic. These places might include videotapes of science programs on television, people in your community, print sources (such as books, magazines, and newspapers), and electronic sources (such as CD-ROMs and Internet sites). The sources of information might be in your school, home, or community.

Evaluate the Sources of Information

Preview your sources of information, and decide whether they are useful. Here are four things to consider.

- *Authority:* Who wrote or developed the information or sponsors the Web site? What are their qualifications?

- *Accuracy:* Are there any obvious errors or inconsistencies in the information? Does the information agree with other reliable sources?

- *Currency:* Is the information up to date? Has recent scientific information been included?

- *Suitability:* Does the information make sense to someone your age? Do you understand it? Is it organized in a way that you understand?

Record and Organize the Information

Identify categories or headings for note taking. Record information, in your own words, under each category or heading, perhaps in point form. If you quote a source, use quotation marks.

Record the sources to show where you got your information. Include the title, author, publisher, page number, and date. For Web sites, record the URL (Web site address).

If necessary, add to your list of questions as you find new information.

Communicate the Information

Choose a format for communication that suits your audience, your purpose, and the information.

COMMUNICATING IN SCIENCE

CREATING DATA TABLES

Data tables are an effective way to record both qualitative and quantitative observations. Making a data table should be one of your first steps when conducting an investigation. You may decide that a data table is enough to communicate your data, or you may decide to use your data to draw a graph. A graph will help you analyze your data. (See "Graphing Data," on page 298, for more information about graphs.)

Sometimes you may use a data table to record your observations in words, as shown below.

Data table for Investigation 7.2

Mineral number	Colour	Streak	Lustre	Hardness	Magnetism	Reaction with vinegar	Cleavage	Name
1	grey-black	reddish brown	metallic					

Sometimes you may use a data table to record the values of the independent variable (the cause) and the dependent variables (the effects), as shown to the left. (Remember that there can be more than one dependent variable in an investigation.)

Follow these guidelines to make a data table:

- Use a ruler to make your table.
- Write a title that describes your data as precisely as you can.
- Include the units of measurements for each variable, when appropriate.
- List the values of the independent variable in the left-hand column of your table.
- List the values of the dependent variable(s) in the column(s) to the right of the independent variable.

Average Monthly Temperatures in Cities A and B

Month	Temperature (°C) in City A	Temperature (°C) in City B
January	-7	-6
February	-6	-6
March	-1	-2
April	6	4
May	12	9
June	17	15

GRAPHING DATA

When you conduct an investigation or do research, you often collect a lot of data. Sometimes the patterns or relationships in the data are difficult to see. For example, look at the data in Table 1.

Table 1 Average Rainfall in Campbell River

Month	Rainfall (mm)
January	142
February	125
March	128
April	73
May	59
June	50
July	40
August	43
September	62
October	154
November	210
December	197

One way to arrange your data so that it is easy to read and understand is to draw a graph. A graph shows numerical data in the form of a diagram. There are three kinds of graphs that are commonly used:

- bar graphs
- line graphs
- circle (pie) graphs.

Each kind of graph has its own special uses. You need to identify which type of graph is best for the data you have collected.

Bar Graphs

A **bar graph** helps you make comparisons and see relationships when one of two variables is in numbers and the other is not. The following bar graph was created from the data in Table 1. It clearly shows the rainfall in different months of the year and makes comparison easy.

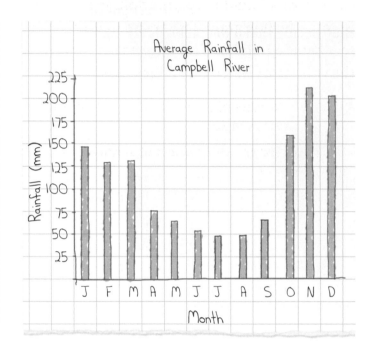

Line Graphs

A **line graph** is useful when you have two variables in numbers. It shows changes in measurement. It helps you decide whether there is a relationship between two sets of numbers: for example, "if this happens, then that happens." **Table 2** gives the number of earthworms found in specific volumes of water in soil. The line graph for these data helps you see that the number of earthworms increases as the volume of water in soil increases.

Table 2 Number of Earthworms per Volume of Water in Soil

Volume of water in soil (mL)	Number of earthworms
0	3
10	4
20	5
30	9
40	22

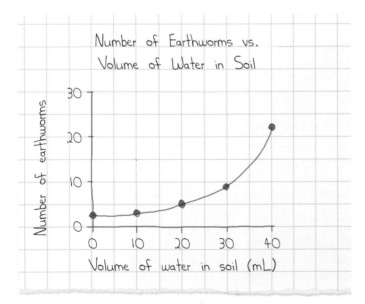

Circle Graphs

A **circle graph** (or pie graph) shows the whole of something divided into all its parts. A circle graph is round and shows how large a share of the circle belongs to different things. You can use circle graphs to see how the different things compare in size or quantity. It is a good way to graph data that are percentages or can be changed to percentages.

WRITING A LAB REPORT

When you design and conduct your own experiment, it is important to report your findings. Other people may want to repeat your experiment, or they may want to use or apply your findings to another situation. Your write-up, or report, should reflect the process of scientific inquiry that you used in your experiment.

> Write the title of your experiment at the top of the page.

> List the question(s) you were trying to answer. This section should be written in sentences.

> Write your hypothesis. It should be a sentence in the form "If … then …."

> Write the materials in a list. Your list should include equipment that will be reused and things that will be used up in the investigation. Give the amount or size, if this is important.

> Describe the procedure using numbered steps. Each step should start on a new line and, if possible, it should start with a verb. Make sure that your steps are clear so that someone else could repeat your experiment and get the same results. Include any safety precautions.
>
> Draw a large diagram with labels to show how you will set up the equipment. Use a ruler for straight lines.

Conductivity of Water

Question

Which type of water — pure water, water with dissolved sugar, or water with dissolved salt — conducts electricity the best?

Hypothesis

If water is very pure, like distilled water with no solutes, then it will conduct electricity better than water with sugar or salt dissolved in it.

Materials

3 clean glass jars	battery holder
distilled water	1 piece of wire, 25 cm long
sugar	2 pieces of wire, each 10 cm long
salt	wire strippers
3 short strips of masking tape	light-bulb holder
pen	small light bulb (such as a flashlight bulb)
2 D-cell batteries	

Procedure

1. Put 250 mL of distilled water in each clean jar. Do not add anything to the first jar. Add 30 mL of salt to the second jar, and mix. Add 30 mL of sugar to the third jar, and mix. Label the jars "pure water," "salt water," and "sugar water."
2. Put the batteries in the holder.
3. Strip the plastic coating off the last centimetre at the ends of all three wires, using the wire strippers.

CAUTION: Always pull the wire strippers away from your body.

4. Attach one end of the 25-cm wire to the knobby end of the battery by tucking it in the battery holder. The other end of the wire should hang free for now.
5. Attach one end of a 10-cm wire to the flat part of the battery. Attach the other end to the clip in the light-bulb holder.

6. Place the light bulb in the holder.
7. Attach one end of the other 10-cm wire to the clip in the light-bulb holder. Let the other end hang free for now.
8. Dip the loose wire ends into the distilled water. Observe whether the light bulb goes on. Record "yes" or "no."
9. Repeat step 8 for the other two types of water.

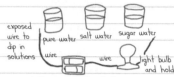

Data and Observations

Type of water	Does the light bulb go on?
distilled water	no
water with salt	yes
water with sugar	no

Analysis

The salt water was the only type of water that turned on the light bulb. Something in the salt must help to conduct electricity. Since the distilled water did not turn on the light bulb, this must mean that it cannot conduct electricity. Something is missing from the distilled water. The sugar water did not conduct electricity either, so it must also be missing the ingredient that helps to conduct electricity.

Conclusion

Pure (distilled) water does not conduct electricity. The hypothesis is not supported by the data, so it is incorrect. Salt water conducts electricity.

Applications

Knowing that salt water conducts electricity might help scientists recover materials from seawater by running electricity through it. Also, I think the water in the human body has salt and other things dissolved in it. It would conduct electricity well, so people should be careful about electricity.

Present your observations in a form that is easily understood. The data should be recorded in one or more tables, with units included. Qualitative observations can be recorded in words or drawings. Observations in words can be in point form.

Interpret and analyze your results. If you have made graphs, include and explain them here. Answer any questions from the student text here. Your answers should include the questions.

A conclusion is a statement that explains the results of an experiment. Your conclusion should refer back to your hypothesis. Was your hypothesis correct, partly correct, or incorrect? Explain how you arrived at your conclusion. This section should be written in sentences.

Describe how the new information you gained from doing your experiment relates to real-life situations. How can this information be used?

GLOSSARY

A

acid a compound that forms a sour-tasting solution, which reacts with metals and can cause serious burns on skin; a solution that is acidic turns blue litmus paper red

acidic a term used to describe a solution that has a value below 7 on the pH scale; the more acidic a solution, the lower its pH value

aftershocks smaller tremors that may occur after an earthquake as pressure in Earth's crust is gradually released; may occur at any time for months after an earthquake

B

base a compound that forms a bitter-tasting solution, which feels slippery, reacts with fats and oils, and can cause serious burns on skin; a solution that is basic turns red litmus paper blue

basic a term used to describe a solution that has a value above 7 on the pH scale; the more basic a solution, the higher its pH value

biodiversity the variety of plant and animal life within an ecosystem; the greater the number of different types of plants and animals, the greater the biodiversity

biological weathering weathering that is aided by living things, such as plants, animals, and micro-organisms

biome large regions of Earth where temperature and precipitation are the same and similar plants and animals are found

biosphere the parts of Earth where life can be found, from mountaintops to the deepest parts of the ocean

boiling point the temperature at which the liquid form of a substance changes to a gas; for example, liquid water changes to water vapour (a gas) at 100°C

C

carnivore a consumer that eats other animals; for example, wolves and orca are carnivores

chemical change a reaction in which the original substance is changed into one or more different substances with different properties; clues that a chemical change has occurred include the production of heat or light, gas bubbles, a colour change, and the formation of new substances

chemical weathering weathering that is caused by a chemical reaction between water, air, or another substance and the materials in rocks

community a group that is made up of two or more populations of different species in an ecosystem

compound a pure substance that is made up of two or more different elements; consists of only one kind of particle

concentration the amount of a substance (the solute) that is dissolved in a given quantity of the substance it is dissolved in (the solvent); the more solute dissolved, the greater the concentration; for example, the concentration of an orange-drink solution depends on the amount of orange-drink crystals dissolved in a given amount of water

condensation the change in the state of a substance from a gas to a liquid; happens when a gas cools and its particles move slower; the opposite of evaporation

consumer an organism, such as an animal, that must obtain its food by eating other organisms in its environment

continental crust the parts of Earth's crust that have continents on them

convergent boundary an area of Earth's crust between plates that are moving toward each other and colliding

crust the "shell" of rock that makes up the hard outermost layer of Earth; "floats" on the inner layers of Earth because it is made of lighter materials than the lower layers

cycle anything that happens over and over again; for example, the seasons of the year and the phases of the moon are both cycles

D

decomposer an organism that gets its food energy by breaking down the final remains of living things, such as dead animals and plants and animal waste; for example, bacteria and fungi are decomposers

delta deposits of sediment in the shape of a triangle at the mouth of a river

density the mass of a substance per unit volume of the substance; expressed as grams per cubic centimeter (g/cm^3) or grams per millilitre (g/mL)

deposition the settling of eroded rock materials on Earth's surface

detrivore an organism that feeds on large bits of dead and decaying plant and animal matter; for example, earthworms, dung beetles, and wolverines are detrivores

dilute a solution that has a low concentration of the dissolved substance (the solute); for example, lemonade with a small amount of dissolved sugar is a more dilute solution than lemonade with a lot of dissolved sugar

dissolve to completely mix one substance (the solute) in another (the solvent) to form a solution; for example, if you add sugar to water, the sugar dissolves in the water

divergent boundary an area of Earth's crust between plates that are moving apart

E

earthquake a vibration of Earth's crust, caused by the sudden release of accumulated energy from plate movement

ecological pyramid a model that shows the effects of the loss of energy in a food chain; at each higher level of the pyramid, the amount of available energy and the number of organisms decrease

ecosystem the network of interactions that link together the living and non-living parts of an environment

element a pure substance that cannot be broken down into any other pure substance; consists of only one kind of particle

emulsion a special kind of suspension that has been treated to prevent the parts of the mixture from separating; for example, homogenized milk is an emulsion

erosion the movement of weathered rock material from one place to another

evaporation the change in the state of a substance from a liquid to a gas; happens when a liquid is heated enough for its particles to break free of each other; the opposite of condensation

extrusive igneous rock igneous rock that is formed on Earth's surface when lava cools

F

fiord a long and narrow inlet of the sea that is formed when valleys become filled with seawater

food chain a model that shows how food energy is passed from one organism to another in a feeding pathway

food web a model that represents several interconnected food chains

fossil rock-like cast, impression, or actual remains of an organism that was buried when it died, before it could decompose

fossil record the history of changes to life on Earth as shown by fossils

freezing (or solidification) the change in the state of a substance from a liquid to a solid; happens when a liquid cools and its particles move more slowly until they settle into fixed positions in a pattern; the opposite of melting

freezing point the temperature at which the liquid form of a substance changes to a solid; for example, liquid water changes to solid ice at 0°C; the freezing point of a substance is the same as its melting point

G

gas a substance with no fixed volume or shape; will fill any container it is in, taking on the container's shape

geologic time scale a time line of the changes to life on Earth

H

habitat the physical space where a certain species lives

herbivore a consumer that eats only plants

hot spot part of the mantle where the temperature is much higher than normal; the magma melts the rock above it and rises to the surface of Earth

I

ice wedging   the widening or splitting of cracks in rocks as rainwater freezes and expands

igneous rock   rock that forms from the cooling and hardening of liquid magma; most of Earth's crust is composed of igneous rock

indigenous knowledge   understandings, values, and beliefs about the natural world that are unique to a particular group or culture who have lived for a very long time in a particular area. This specialized knowledge is passed from generation to generation in the form of stories told, experiences shared, or songs sung by Elders or other people

inner core   the innermost layer of Earth, which is made up of iron and nickel

intrusive igneous rock   igneous rock that is formed when magma cools below Earth's surface

L

lava   magma that is forced out of cracks onto Earth's surface

liquid   a substance with a fixed volume but no fixed shape; takes the shape of the container it is in; particles can move around more freely in a liquid than they can in a solid

M

magma   hot molten rock; cools to form igneous rock

mantle   the layer of Earth between the crust and the outer core; a hot, thick layer of solid and partly melted rock

mass   a measurement of the amount of matter in an object; usually measured in milligrams (mg), grams (g), or kilograms (kg); an object's mass stays constant everywhere in the universe

matter   anything that has mass and volume (occupies space)

mechanical mixture   a mixture in which two or more different parts can be seen with the unaided eye; for example, granola cereal is a mechanical mixture

mechanical weathering   weathering caused by a physical force such as ice, wind, or water

melting   the change in the state of a substance from a solid to a liquid; happens when a solid is heated enough to free its particles from their fixed positions; the opposite of freezing

melting point   the temperature at which the solid form of a substance changes to a liquid; for example, water changes from solid ice to liquid water at 0°C; the melting point of a substance is the same as its freezing point

metamorphic rock   rock that is formed below Earth's surface when heat and pressure cause the characteristics of the existing rock to change

micro-organism   a living thing that is too small to be seen without the help of a microscope; for example, bacteria and some algae are micro-organisms

mid-ocean ridge   the long underwater mountain range that runs through the middle of the oceans

mineral   pure, naturally occurring substance that is found in Earth's crust; the building block of rock; for example, diamonds, graphite, and talc are minerals

mixture   any substance that contains two or more pure substances and therefore has two or more kinds of particles; properties of mixtures can be different in different samples

mountain   a landmass that rises significantly from the surrounding level of Earth's surface

N

neutral   neither an acid nor a base; on the pH scale, a neutral substance or solution has a value of 7

niche   the way that an organism fits into an ecosystem, in terms of where it lives, how it obtains its food, and how it interacts with other organisms

non-reversible change   a change in a substance that cannot be reversed; for example, wood sawed into pieces cannot be put together to form the original piece of wood again

O

oceanic crust   the parts of Earth's crust that have only ocean floor on them

omnivore   a consumer that eats both plants and animals

organism   a living thing, such as a plant or an animal

outer core   a dense, hot layer of Earth between the mantle and the inner core; made up of mostly liquid iron and some nickel

P

Pangaea the name of a hypothetical supercontinent proposed by Alfred Wegener

particle model a scientific model that describes matter as made up of tiny particles with spaces between the particles; the particles are always moving, and adding heat makes them move faster

pH a scale that measures the acidity of substances; has numbers from 0 (strongly acidic) to 7 (neutral) to 14 (strongly basic)

photosynthesis the process in which the Sun's energy is used by plants to produce simple sugars from carbon dioxide and water; oxygen is released in this process

physical change a change in the properties of a substance, such as its form or state; the substance itself does not change; for example, a piece of wood cut into pieces is still wood, and melted wax is still wax

plain a level area of land; usually in the interior of a continent

plateau a large area of high, fairly flat ground

plate tectonics the theory that the surface of Earth consists of approximately a dozen large plates that are continually moving

population all the members of one particular species living in a given area

predator an organism that hunts another living thing for food

preserve to maintain something in its existing state; for example, you can help to preserve an ecosystem by working to keep it in its current state

prey an organism that is hunted by a predator

producer an organism that can make its own food from non-living materials

property a characteristic of a material that can be observed (colour or lustre) or determined through simple tests and measurements (density or melting point)

pure substance a substance that contains only one kind of particle throughout and therefore always has the same properties; there are two kinds of pure substances: elements and compounds

R

reversible change a change in a substance that can be reversed; for example, melted wax can be cooled to form solid wax again

rock cycle the changing of igneous, sedimentary, and metamorphic rocks from one into another over a long period of time

S

saturated a solution in which no more of one substance (the solute) can be dissolved in another substance (the solvent); for example, when you cannot dissolve any more drink crystals in water, the solution is saturated

sediment solid particles (such as rock particles, clay, mud, sand, gravel, and boulders) that are carried by moving water and gradually settle onto the floor of a lake or ocean

sedimentary rock rock that is formed by the breaking down, depositing, compacting, and cementing of sediment

seismic wave an energy wave that is caused by an earthquake

solid a substance with a fixed shape and a fixed volume; the particles in a solid only move a little—they vibrate back and forth but remain in a fixed position in a pattern

solubility the ability of a substance (the solute) to dissolve in another substance (the solvent); temperature plays an important role in solubility; for example, you can dissolve more orange-drink crystals in warm water than in cold water

solution a mixture of two or more substances that appears to be made up of only one substance; for example, clear apple juice (a liquid), clean air (a gas), and stainless steel (a solid mixture of metals) are all solutions

species a term used to describe each different kind of organism; for example, all dogs (from toy poodles to great Danes) belong to the same species because they can mate and reproduce fertile offspring; cats belong to a different species than dogs

state a property describing whether a substance is a solid, a liquid, or a gas; for example, water can be found as a solid (ice), a liquid (water), or a gas (water vapour in the air)

stewardship taking personal responsibility for something; for example, by caretaking in an ecosystem

subduction zone an area of Earth's crust where one plate is sinking below another

sublimation the change in state of a substance from a solid to a gas without first becoming a liquid; happens when particles of a solid gain enough energy to break completely away from the other particles, forming a gas

supersaturated a solution that is more than saturated; contains more of the dissolved substance (the solute) than would normally be found in a saturated solution

suspension a cloudy mixture in which clumps of a solid or droplets of a liquid are scattered throughout a liquid or gas; for example, muddy water is a suspension

sustainability the ability of ecosystems to bear the impact of the human population over a long period of time, through the replacement of resources and the recycling of waste

T

transform fault boundary an area of Earth's crust between plates that are slipping past each other

tsunami an ocean wave that is caused by an earthquake or an underwater volcano

U

unsaturated a solution in which more of one substance (the solute) can still be dissolved in another substance (the solvent); for example, when you can still dissolve drink crystals in water, the solution is unsaturated

V

valley a low region of land between hills or mountains

volcano any opening in Earth's crust through which molten rock and other materials erupt

volume a measure of the amount of space that is occupied by matter; the volume of a liquid is generally measured in millilitres (mL) or litres (L); the volume of a solid is usually measured in cubic centimetres (cm^3); 1 cm^3 is the same as 1 mL, and 1000 cm^3 equals 1 L

W

weathering the process that slowly breaks down natural materials (such as rocks and boulders) into smaller pieces

INDEX

Sugar, 29, 137, 156
Sun, energy from, 29, 41, 52, 53, 64
Sunlight, 7, 15, 17
Supersaturation, 156–157
Survival
  of animals, 15
  chemical changes and, 123
  of humans, 59
  interactions and, 3
  of organisms, 7, 16
  of plants, 15
  short-term changes in temperature
    and, 19
Suspensions, 145, 147
Sustainability, 60, 61

T

Ta-gil, 254
Tahltan First Nation, 261
Temperature, 18–19, 64
  measurement of, 288
Transform fault boundaries, 243, 247
Tree diagrams, 290
Trenches, 226, 227, 229
Tsunamis, 250–251, 254
Tubeworms, 40

U

Unsaturated solutions, 155–156

V

Valleys, 205, 207, 236, 238
Vancouver Island, 195, 201
Variables, 270–271
Venn diagrams, 292
Verne, Jules, 220
Volcanoes, 211, 228, 230, 236,
  255–261
Volume, 101–103
  measurement of, 286–287

W

Wackernagel, Mathis, 82
Wastes, recycling of, 59, 60
Wastewater, 51, 65
Water, 15, 17–18
  as compound, 141
  cycle, 52–54
  dissolving of rock materials by, 201
  ecosystems, 17
  erosion by, 205–208
  evaporation of, 53
  flow of, 13
  fresh, 13, 18
  ice becoming, 94
  nutrients in, 44
  pollution of, 65, 66
  as pure substance, 137
  salt, 13, 18
  use by plants, 29
  vapour, 53, 94, 119
  and weathering of rock, 200
Weather, 198
Weathering, 198
  biological, 202–203
  chemical, 201–202
  mechanical, 199–200
Wegener, Alfred, 221–226
Wilson, J. Tuzo, 243, 258
Wind
  erosion by, 204–205
  and weathering of rock, 199

Y

Yeast, 21, 66

PHOTO CREDITS

Cover

Photodisc Blue/Photodisc Collection/Getty Images

Preface

p. x left to right Lester Lefkowitz/CORBIS/MAGMA, CORBIS/MAGMA, Ed Young/CORBIS/MAGMA, Bettmann/CORBIS/MAGMA; p. 1 Al Harvey/ www.slidefarm.com.

Table of Contents

p. vii Wendy Shymanski; p. viii © Ken Laffal/Index Stock Imagery; p. ix Alan Sirulnikoff.

Unit A

Unit A Opener pp 2–3 Wendy Shymanski; p. 4 © Inga Spence/Index Stock Imagery; p. 5 Wendy Shymanski; p. 7 Wendy Shymanski; p. 8 left Dave Starrett, right Al Harvey/ www.slidefarm.com; p. 11 left Dick Hemingway, right Fran Gealer/picturearts/ First Light; p. 13 Earth Imaging/ Getty Images; p. 14 John Wanderer; p. 15 Osoyoos Desert Society; p. 16 D. Taylor/IVY IMAGES; p. 17 top © Brandon D. Cole/CORBIS/MAGMA, middle D. Taylor/IVY IMAGES, bottom Carmen Dawkins; p. 18 top Dean Van't Schip, bottom Al Harvey/ www.slidefarm.com; p. 19 left Dean Van't Schip, right Al Harvey/www.slidefarm.com; p. 20 left Dave Starrett, right M. Kalab/Visuals Unlimited; p. 22 left by Fred Davis, Haida artist. Courtesy of the Spiritwrestler Gallery. Photo by Kenji Nagai., right Carmen Dawkins; p. 23 Allan & Sandy Carey/IVY IMAGES; p. 26 top Earth Imaging/ Getty Images, bottom left Carmen Dawkins, bottom right Allan & Sandy Carey/IVY IMAGES; p. 28 Victoria Hurst/ First Light; p. 30 top left Photodisc/Getty Images, top right Al Harvey/www.slidefarm.com, bottom left Al Harvey/www.slidefarm.com, bottom right Dean Van't Schip; p. 31 top left Photodisc/ Getty Images; top right Al Harvey/ www.slidefarm.com, middle left G. Peck/Ivy Images, middle right © Raymond Gehman/ CORBIS/MAGMA, bottom Breck P. Kent/Animals Animals, bottom right Bob Semple; p. 32 © Kevin King, Ecoscene/CORBIS/MAGMA; p. 36 left Dave Starrett, right Anne Bradley; p. 38 Anne Bradley; p. 39 John Gerlach/ Visuals Unlimited; p. 40 Science VU/Visuals Unlimited; p. 44 NASA; p. 45 top Stephen Sharnoff/Visuals Unlimited, bottom Sherman Thomson/Visuals Unlimited; p. 47 Al Harvey/www.slidefarm.com; p. 48 Dave Starrett; p. 50 Dick Hemingway; p. 51 Courtesy of University of British Columbia. Photo by Martin Dee; p. 53 Al Harvey/

www.slidefarm.com; p. 58 Suzanne L. Collins/ Photo Researchers, Inc.; p. 60 David Neel; p. 61 Al Harvey/ www.slidefarm.com; p. 62 top Al Harvey/ www.slidefarm.com, bottom Al Harvey/www.slidefarm.com; p. 63 top Peter Battistoni/Vancouver Sun, bottom left Dick Hemingway, bottom right Dick Hemingway; p. 64 Courtesy of International Institute for Sustainable Development. Photo by Neil Ford; p. 65 CORBIS/ MAGMA; p. 66 Dave Starrett; p. 67 Lyle Ottenbreit; p. 68 left Victor Last/Geographical Visual Aids, right Victor Last/Geographical Visual Aids; p. 69 top left Courtesy of Burrowing Owl Winery, top right Al Harvey/ www.slidefarm.com, bottom left Al Harvey/ www.slidefarm.com, bottom right © Paul A. Souders/ CORBIS/MAGMA; p. 70 Brian Sprout/Vancouver Sun; p. 71 Al Harvey/www.slidefarm.com; p. 72 Lyle Ottenbreit; p. 75 Courtesy of International Institute for Sustainable Development. Photo by Graham Ashford; p. 76 top to bottom © Inc, Grant Heilman Photography/ Index Stock Imagery, © Diaphor Agency/Index Stock Imagery, © David Samuel Robbins/CORBIS/MAGMA, Peter Van Rhijn/Superstock, G. Peck/IVY IMAGES, Steve Maslowski/Visuals Unlimited; p. 77 Joe McDonald/ Visuals Unlimited; p. 78 Rick Blacklaws; p. 79 Dave Saunders; p. 80 top left Phil Hallinan, top right Al Harvey/www.slidefarm.com, bottom left Blair Acton, bottom right Adrian Dorst; p. 81 top left Courtesy of Bamfield Huu-ay-aht Community Abalone Project, top right Al Harvey/www.slidefarm.com, bottom left Mike Mackintosh, bottom right Robert Leon; p. 83 Courtesy of David Suzuki Foundation; p. 84 top left Al Harvey/www.slidefarm.com, top middle Al Harvey/ www.slidefarm.com, top right © Inc, Grant Heilman Photography/Index Stock Imagery, bottom left Wendy Shymanski, bottom middle Blair Acton, bottom right Dave Saunders; p. 86 Al Harvey/ www.slidefarm.com.

Unit B

Unit B opener: p. 88–89 © Ken Laffal/Index Stock Imagery; p. 90 © Yuichi Takasaka; p. 92 D. Trask/ Ivy Images; p. 93 © Paul Seheult; Eye Ubiquitous/CORBIS/MAGMA; p. 94 left Dave Starrett, right Alan Sirulnikoff; p. 95 Dave Starrett; p. 96 Dave Starrett; p. 98 top left © Tom Walker/Index Stock Imagery, top right © Kevin R. Morris/CORBIS/MAGMA, middle Jeff Daly/Visuals Unlimited, bottom Courtesy of Applied Robotics, Inc.; p. 99 © Jose Luis Pelaez, Inc./CORBIS/ MAGMA; p. 100 left & right Dave Starrett, middle

© Tom Stewart/ CORBIS/MAGMA; p. 104 left
Dave Starrett, right Anne Bradley; p. 105 Lyle Ottenbreit;
p. 106 Dave Starrett; p. 107 top © Natalie Fobes/CORBIS/
MAGMA, bottom Courtesy of Keet Gooshi Tours; p. 108
Dave Starrett; p. 109 © Joel W. Rogers/CORBIS/MAGMA;
p. 110 Dave Starrett; p. 111 top & bottom Ray Boudreau;
p. 112 Larry Stepanowicz/Visuals Unlimited; p. 113 left
D. Trask/IVY IMAGES, right © Paul Seheult; Eye
Ubiquitous/CORBIS/MAGMA, bottom left Dave Starrett;
p. 114 Anne Bradley; p. 115 left Courtesy of Keet Gooshi
Tours, right Photodisc; p. 116 © Tom Stewart/CORBIS/
MAGMA; p. 118 top four Dave Starrett, bottom Anne
Bradley; p. 119 © Omni Photo Communications Inc./
Index Stock Imagery; p. 120 left Anne Bradley, right
Lawrence Manning/CORBIS/ MAGMA; p. 121 left
© ThinkStock LLC/Index Stock Imagery, right
Anne Bradley; p. 122 left Tom Bean/CORBIS/MAGMA,
right © LLC, FogStock/Index Stock Imagery; p. 124 top left
Bill Ivy/IVY IMAGES, top middle Dave Starrett, top right
© FoodCollection/Index Stock Imagery, bottom left
W. Fraser/Ivy Images, bottom middle Dave Starrett, bottom
right Anne Bradley; p. 125 Eyewire/Getty Images; p. 126
Dave Starrett; p. 127 Lyle Ottenbreit; p. 128 left
Noel Hendrickson/Masterfile, right Bill Ivy/IVY IMAGES;
p. 129 Bill Ivy/IVY IMAGES; p. 130 left Dick Hemingway,
right Al Harvey/www.slidefarm.com; p. 131 top
© Roger Ressmeyer/CORBIS/MAGMA, bottom left
© William Swartz/Index Stock Imagery, bottom right
Loren Winters/Visuals Unlimited; p. 132 left © ThinkStock
LLC/Index Stock Imagery, right Tom Bean/CORBIS/
MAGMA; p. 133 top left Anne Bradley, top right
© LLC, FogStock/Index Stock Imagery, bottom left
Noel Hendrickson/Masterfile, bottom right Al Harvey/
www.slidefarm.com; p. 134 RDF/Visuals Unlimited; p. 135
top © CORBIS/MAGMA, bottom Jeff Daly/Visuals
Unlimited; p. 136 © Dennis Marsico/CORBIS/MAGMA;
p. 137 top © Charles O'Rear/CORBIS/MAGMA, bottom
Anne Bradley; p. 138 Dave Starrett; p. 139 top Gary
Randall/Visuals Unlimited, bottom Dave Starrett; p. 140
Anne Bradley; p. 141 Anne Bradley; p. 142 top
Anne Bradley, bottom left Andrew Lambert
Photography/Science Photo Library, bottom right
Andrew Lambert Photography/Science Photo Library;
p. 143 Anne Bradley; p. 144 Dave Starrett; p. 145 top
Photodisc/Getty Images, bottom David Barr; p. 146
Anne Bradley; p. 151 Dave Starrett; p. 153 Dave Starrett;
p. 154 Ray Boudreau; p. 157 left ©1990 Richard Megna
Fundamental Photographs, NYC, right ©1990
Richard Megna Fundamental Photographs, NYC; p. 158
Dave Starrett; p. 161 top Dave Starrett, bottom left
Andrew Lambert Photography/Science Photo Library,
bottom right Andrew Lambert Photography/Science Photo

Library; p.163 left & right Dave Starrett; p. 164 all
Anne Bradley; p. 165 Dave Starrett; p. 166 D. Trask/
IVY IMAGES; p. 168 Anne Bradley; p. 169 top left
Photodisc/Getty Images, middle David Barr, top
right Anne Bradley, bottom left Dave Starrett; p. 171
© Natalie Fobes/CORBIS/MAGMA; p. 172 Dave Starrett;
p. 173 Lyle Ottenbreit; p. 174 left Lyle Ottenbreit, right
Anne Bradley.

Unit C

Unit C opener: p. 176–177 Alan Sirulnikoff; p. 178
Mark Schneider/Visuals Unlimited; p. 179 left
Albert Copley/Visuals Unlimited, right Dave Starrett;
p. 180 top left Ken Lucas/Visuals Unlimited, top right
Ivy Images, bottom left & bottom right Dave Starrett;
p. 181 left Mark A. Schneider/Photo Researchers, Inc.,
middle Albert Copley/Visuals Unlimited, right
Charles D. Winters/Photo Researchers, Inc.; p. 182 top left
Ken Lucas/Visuals Unlimited, top right Mark Schneider/
Visuals Unlimited, bottom left ©Charles Winters/Photo
Researchers, Inc, bottom right Carolina Biological/Visuals
Unlimited; p. 183 top Anne Bradley, bottom Dave Starrett;
p. 184 left & right Dave Starrett; p. 187 top Albert Copley/
Visuals Unlimited, middle Dave Starrett, bottom
Anne Bradley; p. 188 Al Harvey/www.slidefarm.com; p. 189
top left Wally Eberhart/Visuals Unlimited, top right
Doug Sokell/Visuals Unlimited, bottom Dave Starrett;
p. 190 CCRS; p. 191 top left Wally Eberhart/Visuals
Unlimited, top middle Wally Eberhart/Visuals Unlimited,
top right Wally Eberhart/Visuals Unlimited, bottom
Al Harvey/www.slidefarm.com; p. 192 clockwise from top
right Wally Eberhart/Visuals Unlimited, Mark Schneider/
Visuals Unlimited, Deborah Long/Visuals Unlimited,
Wally Eberhart/Visuals Unlimited, Wally Eberhart/Visuals
Unlimited, Dave Starrett, Wally Eberhart/Visuals
Unlimited, Wally Eberhart/Visuals Unlimited; p. 193
left Albert Copley/Visuals Unlimited, middle Joyce
Photographics/Photo Researchers, Inc., right
Doug Martin/Photo Researchers, Inc.; p. 194 top
Charles Helm, bottom Ken Lucas/Visuals Unlimited;
p. 195 Ken Lucas/Visuals Unlimited; p. 196 top to bottom
© Neil Rabinowitz/ CORBIS/MAGMA, Barbara
Strnadova/Photo Researchers, Inc., © Kevin Schafer/
CORBIS/MAGMA, © James L. Amos/CORBIS/
MAGMA, Ken Lucas/Visuals Unlimited; p. 197 left
Alan Sirulnikoff/Science Photo Library, right Dept of
Paleobiology, Smithsonian Institute; p. 198 Marli
Miller/Visuals Unlimited; p. 199 Bill Lowry/IVY IMAGES;
p. 200 top Al Harvey/www.slidefarm.com, middle
Martin Miller/Visuals Unlimited, bottom Dave Starrett;
p. 201 Terry Parker; p. 202 top John D. Cunningham/
Visuals Unlimited, bottom Al Harvey/www.slidefarm.com;

p. 203 Janet Dwyer/First Light; p. 204 Science VU/Visuals Unlimited; p. 205 top left Glenbow Archives NA-2658.85, top right NASA, bottom LANDSAT 7 ETM+ data 2000. Received by the Canada Centre for Remote Sensing(CCRS). Processed and distributed by RADARSAT International under licence from CCRS; p. 206 top Victor Last/Geographical Visual Aids, bottom Bill Lowry/IVY IMAGES; p. 207 top © David Muench/CORBIS/MAGMA, bottom Al Harvey/www.slidefarm.com; p. 208 top left © Natalie Fobes/CORBIS/MAGMA, top right © Gunter Marx Photography/CORBIS/ MAGMA, far left Dave Starrett; p. 211 © Jim Sugar/CORBIS/MAGMA; p. 212 left Dave Starrett, right Ray Boudreau; p. 213 Ray Boudreau; p. 214 top (crystal) © Carolina Biological/Visuals Unlimited, (lustre) Charles D. Winters/Photo Researchers, Inc., (magnetism) Anne Bradley, (hardness, colour, & reaction) Dave Starrett, middle left to right © Jim Sugar/CORBIS/MAGMA, Al Harvey/www.slidefarm.com, Doug Martin/Photo Researchers, Inc., bottom Ken Lucas/Visuals Unlimited; p. 215 top left to right Bill Lowry/IVY IMAGES, Terry Parker, Janet Dwyer/First Light, middle left Science VU/Visuals Unlimited, middle right Victor Last/Geographical Visual Aids, bottom left Glenbow Archives NA-2658.85, bottom right Al Harvey/www.slidefarm.com; p. 216 left Ken Lucas/Visuals Unlimited, right Al Harvey/www.slidefarm.com; p. 217 bottom Dave Starrett; p. 218 Al Harvey/www.slidefarm.com; p. 220 all Dave Starrett; p. 222 left Albert Copley/Visuals Unlimited, right Ken Lucas/Visuals Unlimited; p. 225 Dave Starrett; p. 226 Tom van Sant/Photo Researchers, Inc.; p. 230 far left Dave Starrett, left © Mohammad Berno/CORBIS/

MAGMA, right Doug Sokell/Visuals Unlimited; p. 233 left Woods Hole Oceanographic Institute, middle Patrick Morrow, right NEPTUNE Canada; p. 234 Tom van Sant/Photo Researchers, Inc.; p. 235 Woods Hole Oceanographic Institute; p. 236 © Jim Sugar/CORBIS/MAGMA; p. 237 Dr. Ken Macdonald/SPL/Photo Researchers, Inc.; p. 238 Simon Fraser/Photo Researchers, Inc.; p. 242 Pegasus/Visuals Unlimited; p. 243 Science VU/Visuals Unlimited; p. 244 Dave Starrett; p. 248 Albert Copley/Visuals Unlimited; p. 250 © Patrick Robert/CORBIS SYGMA/MAGMA; p. 251 Charles Ford; p. 252 © Peter Yates/CORBIS/MAGMA; p. 254 by Tim Paul. Courtesy of Spiritwrestler Gallery. Photo by Kenji Nagai; p. 255 © Roger Ressmeyer/CORBIS/MAGMA; p. 256 © Bettmann/CORBIS/MAGMA; p. 257 © Bruce Ely/The Oregonian/CORBIS/MAGMA; p. 258 © Michael T. Sedam/CORBIS/MAGMA; p. 260 © Gunter Marx Photography/CORBIS/MAGMA; p. 261 Wally Eberhart/Visuals Unlimited; p. 262 left Simon Fraser/Photo Researchers, Inc., middle Pegasus/Visuals Unlimited, right Science VU/Visuals Unlimited.

Skills Handbook
p. 266 Lyle Ottenbreit; p. 268 top & left Dave Starrett, bottom right Ray Boudreau; p. 269 Dave Starrett; p. 270 Ray Boudreau; p. 271 Dave Starrett; p. 272 Dave Starrett; p. 274 left Courtesy of Boreal, right Dave Starrett; p. 275 right Rick Fischer/ Masterfile; p. 282 Lyle Ottenbreit; p. 288 top left Boreal, bottom left Richard L. Carlton/Visuals Unlimited, right Dave Starrett; p. 296 left to right Todd Ryoji, Omni Photo Communications/Index Stock, Dave Starrett, Dick Hemingway.

TEXT CREDITS

Unit A
p. 70 Material reprinted with the express permission of: "Pacific Newspaper Group", a CanWest Partnership; p. 83 Courtesy of David Suzuki Foundation.

Unit B
p. 143 ©2004, Flinn Scientific, Inc. All Rights Reserved. Reproduced for one-time use with permission from Flinn Scientific, Inc., Batavia, Illinois, U.S.A. No part of this material may be reproduced or transmitted in any form or by any means, electronic or mechanical, including, but not limited to photocopy, recording, or any information storage and retrieval system, without permission in writing from Flinn Scientific, Inc.

Unit C
p. 254 "Native Myths Shed Light on BC's Past", part four of the "Big Ideas of 2003" series, National Post, January 2, 2004. Material reprinted with the express permission of: "Pacific Newspaper Group", a CanWest Partnership.